W9-AQM-211

An Introduction to Controlled Thermonuclear Fusion

An Introduction to Controlled Thermonuclear Fusion

M. O. Hagler
M. Kristiansen
Texas Tech University

Lexington Books
D. C. Heath and Company
Lexington, Massachusetts
Toronto

Sci
QC
791.73
H33

Library of Congress Cataloging in Publication Data
Hagler, M. O
    An introduction to controlled thermonuclear fusion.
    Bibliography: p.
    Includes index.
    1. Controlled fusion. 2. Fusion reactors. I. Kristiansen, Magne,
1932- joint author. II. Title.
QC791.H33   621.48'4   76-33596
ISBN 0-669-99119-8

Copyright © 1977 by D. C. Heath and Company

All rights reserved. No part of this publication may be reproduced or transmitted in any form or by any means, electronic or mechanical, including photocopy, recording, or any information storage or retrieval system, without permission in writing from the publisher.

Published simultaneously in Canada

Printed in the United States of America

International Standard Book Number: 0-669-99119-8

Library of Congress Catalog Card Number: 74-33596

To Shirlene and Aud

459379

Contents

List of Figures

List of Tables

Preface

This book is intended to help those struggling through the morass of plasma physics, nuclear physics, and materials science which makes up controlled thermonuclear fusion to gain an overall perspective in a relatively painless way. The book is not a text on plasma physics, nuclear physics, or materials science, however, and no real prerequisite knowledge is required in these areas in order to use it. That is, the book is more or less self-contained. Even so, we have not hesitated to introduce and explain relatively sophisticated concepts when they are needed to tell the story about how controlled thermonuclear fusion is supposed to work. We try very hard to develop clear intuitive pictures of the various processes and approaches in controlled thermonuclear fusion starting from concepts with which most engineers and physical scientists are familiar ($F = ma$, $p = nkT$, Gauss's law, $F = q (E + v \times B)$, Ampere's law, and the like). Our book could therefore be useful to, for example, (1) students taking an introductory plasma course who want to see what plasmas have to do with fusion, (2) engineers and physical scientists who have recently come into the fusion field to work in a particular problem area and who want to see how this area fits into the overall effort, and (3) those who have been working in fusion for some time, perhaps, but who have not been able to take time out to survey the activity in the field in order to see the ideas behind what all the other people in fusion are doing. In short, the book is intended either for self study or as a supplementary text for undergraduate or graduate courses related to fusion. Having students in an introductory course study as outside reading the relatively simple treatment of Tokomaks and to discuss it briefly in class, for example, should give them a much better idea about MHD equilibrium and stability, neoclassical and anomalous diffusion, trapped particle modes, and scaling in Tokamaks than would otherwise be possible. The book can also be complementary to most plasma texts in that the development of some important plasma physics concepts proceeds along a considerably more physical and less abstract route than is customary. Note, for example, the discussion of drift waves in the Tokamak sections. Exposure to more than one approach is almost always useful.

In summary, it is clear that this book is not intended to transform its users into experts on controlled fusion. Rather it is intended to provide a simple, hopefully coherent, introduction accessible to workers with diverse backgrounds.

We have not tried to provide a history of the development of controlled fusion. Although this subject is a fascinating one, we feel that from a pedagogic viewpoint, history is not always the best teacher. Amasa S.

xvii

Bishop's book, listed in the annotated bibliography, discusses the history of controlled fusion to 1958. After this date, one must rely on scattered review articles and the like.

Superscripted, lower-case letters refer to special notes at the end of each chapter. Also note that rationalized MKS units have been used unless otherwise noted.

We gladly acknowledge the cooperation of those who provided information, and in particular, drawings for our use here. We also appreciate those who have read and commented on various parts of the manuscript. Finally, we award red badges to Vicky Todd, Mary Garcia, Lyna Cattaneo, and Susan Godwin who, seemingly without fear or trepidation, attacked piles of what appeared to be crumpled scrap paper with quasi-legible scribbling on front, back, margins, between the lines . . . , and successfully transformed them into a typewritten copy of the manuscript.

**Part I
General Considerations**

1

Background Theory

Introduction

A promising alternative to obtaining power from chemical reactions (burning fossil fuels) is obtaining power from nuclear reactions (burning nuclear fuels). The reactions of interest can be classified as either *fission* or *fusion*. In fission reactions, a heavy nucleus splits into two lighter ones plus two or three neutrons with high kinetic energy. In fusion reactions, two light nuclei join to form a heavier and a lighter one with high kinetic energy. This kinetic energy must, in both cases (fission and fusion), be converted to some useful form such as electricity.

Fission reactions are the basis of all present day nuclear power plants and of the breeder reactors presently under development. We discuss, on the other hand, some of the principles of nuclear *fusion* reactor design. Although there are still many technical problems that must be overcome before a nuclear fusion reactor can actually be designed and constructed, the interest in producing electric power from controlled thermonuclear fusion reactions is increasing. This interest, reflected by publication of several recent review papers [1–5] on controlled thermonuclear fusion, is due in part to the promise such reactors hold for safely producing relatively cheap electric power. Several recent encouraging developments in fusion research, after many years of disappointment, have been the main reason for this renewed interest, however. We refer the reader to these review papers [1–5] for further discussions of the potential advantages of fusion power.

Fission, Fusion, Mass Defect, and Energy Release

We begin our consideration of how fission and fusion processes release energy by recalling that the nuclei of atoms consist basically of protons and neutrons and that the mass of a free (isolated) proton is 1.007825 amu and that of a free (isolated) neutron 1.008665 amu.[a] If we were to assemble a nucleus from a group of isolated protons and neutrons and then determine its mass, we would find that it would, in general, be different from the sum of the masses of the (isolated) component protons and neutrons. For example, a He^4 nucleus (an α-particle) consists of two neutrons and two protons and has a mass of 4.00260 amu, but the mass of two free neutrons and two free protons totals 4.03298. The mass of any nucleus (except hydrogen) turns out to be less than the total mass of its proton and neutron components. The difference is called the *mass defect*. Figure 1–1 shows the mass defect per nucleon (a proton or a neutron) plotted versus the mass number (the total number of neutrons and protons in the nucleus). The mass defect per nucleon for a particular nucleus is simply the mass defect of the nucleus divided by the mass number of the nucleus. Note that the maximum mass defect per nucleon occurs for mass numbers near 60. Thus we can increase the mass defect per nucleon either by fusing light elements or by fissioning heavy elements. So what?

Well, we know from the law of equivalence of mass and energy that if mass disappears, it must change into energy. Thus since both fusing light nuclei and fissioning heavy nuclei decreases the average mass per nucleon (that is, increases the mass defect), then both processes release energy that has been stored in the nucleus. The energy released appears primarily as kinetic energy of the reaction products.

The Coulomb Barrier

A fundamental problem in fusing two nuclei is that they repel each other because they are both positively charged. If the nuclei meet head on with high enough energy, however, they can get close enough together so that the short range nuclear forces (which hold all nuclei together despite their net positive charges) overcome the Coulomb repulsion and fusion occurs. Roughly speaking, the nuclear attractive forces begin to dominate when the nuclei "touch." The term "touch" is used loosely since the nucleus, a quantum mechanical particle, does not have a clearly defined boundary. Nevertheless, the radius R of a nucleus is approximately given by[b]

$$R = 1.5 \times 10^{-15}A^{1/3} \text{ meters}$$

where A is the atomic mass number of the nucleus. Thus, when a nucleus with atomic number Z_1 and radius R_1 just touches a nucleus with atomic number Z_2 and radius R_2, the potential energy due to Coulomb repulsion is

$$(U_{\text{repulsion}})_{\text{max}} = \frac{(Z_1e)(Z_2e)}{4\pi\epsilon_0(R_1 + R_2)} \tag{1.1}$$

For isotopes of hydrogen, $Z_1 = Z_2 = 1$ and $R_{1,2} \approx 2 \times 10^{-15}$m. Thus,

$$(U_{\text{repulsion}})_{\text{max}} \approx 360 \text{ keV}^c$$

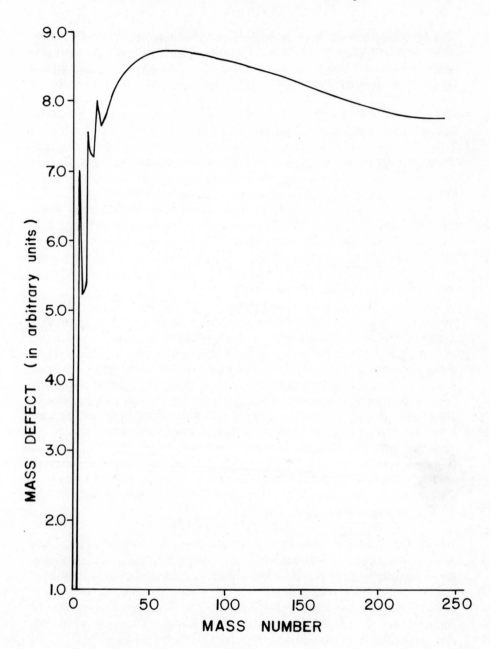

Figure 1–1. Mass Defect versus Mass Number.

Notice that elements with small atomic numbers make $(U_{repulsion})_{max}$ small. From Figure 1–1, it is also clear that nuclei with lower mass release more energy per nucleon in a fusion reaction than do nuclei with higher mass. Thus, both of these factors favor light nuclei for fusion.

Possible Fusion Reactions of Interest

Some of the reactions of interest for controlled thermonuclear fusion are:[d]

$$D + D \rightarrow He^3 + n + 3.27 \text{ MeV}$$
$$D + D \rightarrow T + H + 4.05 \text{ MeV}$$

almost equal probability of occurrence

$$D + T \rightarrow He^4 + n + 17.58 \text{ MeV}$$ (1.2)

$$D + He^3 \rightarrow He^4 + H + 18.34 \text{ MeV}$$

$$H + Li^6 \rightarrow He^4 + He^3 + 4 \text{ MeV}$$

where D (deuterium) is the isotope of hydrogen with atomic mass number 2 (one proton and one neutron in the nucleus) and T (tritium) is the isotope of hydrogen with atomic mass number 3. One gallon of seawater contains enough deuterium to release the energy approximately equivalent to that released in burning 300 gallons of gasoline. Since the cost of separating out the deuterium is quite low, fuel costs in a D-D reactor would be negligible. Deuterium occurs in nature in the ratio of 1/6500 relative to regular hydrogen. Tritium, on the other hand, is radioactive with a half life of 12.4 years and thus is not found in significant quantities in nature. Any tritium fuel must thus be man-made.

Fusion in Colliding Beams of Particles

By definition, an electron-volt (eV) is the amount of energy a particle with an electric charge equal to that of an electron gains when it falls through a potential difference of one volt. Thus, nuclei could be accelerated to energies exceeding those required to overcome the Coulomb barrier[e] (~360 keV) by standard particle accelerators—Van de Graaff generators can accelerate particles to more than 3 MeV, for example. Thus, we might propose firing beams of charged MeV particles at each other to achieve fusion [6]. Unfortunately, the probability for particles being scattered out of the beam and lost is several thousand times greater than the probability for two nuclei fusing. Since particles scattered out of the beam represent an energy loss, it is impossible to recover as much energy from fusion reactions produced in this manner as it takes to accelerate all the particles to the energy required to overcome the Coulomb barrier.[f] It can also be shown that the energy density produced by such a reactor scheme would be much too low for practical interest in large scale power production even if the particle scattering were not a problem.[g]

Fusion of Particles with Random Thermal Motion

The most commonly proposed technique for overcoming the difficulties of the large Coulomb scattering probability is to confine a collection of nuclei and heat them to a high temperature. In this way, a particle undergoing rapid random thermal motion in a group of thermal particles has many more chances to collide and fuse than does a similar particle moving linearly along with a group of particles in a beam. This technique is called thermonuclear fusion.

Given a group of particles with temperature T, the average thermal energy per particle is $3/2\ kT$ where T is the temperature in °K and k is the Boltzmann constant.[h] We would certainly expect to get an appreciable number of fusion reactions if the average particle energy is 360 keV or more. This corresponds to a temperature of about 4×10^9 °K. Actually somewhat lower temperatures are adequate because, even at reduced temperatures, many of the particles in the group have energies large enough to overcome the Coulomb barrier. Because the energy released by each fusion reaction is relatively large, only a small fraction of the particles needs to fuse in order to release a substantial amount of energy.[i] It is clear, however, that no material walls could withstand even the considerably reduced temperature since most materials melt if $T > 3500$°K. Actually of more importance is the fact that the material walls quickly cool the nuclei to a temperature so low that no fusion reactions would occur. Thus some means other than a material container (bottle) must be used to confine the nuclei. In the next section, we begin a discussion of magnetic confinement. Inertial confinement, important in laser fusion schemes, is briefly discussed in a later section.

Magnetic Confinement of a Single Charged Particle

A single charged particle moving in a constant, uniform, magnetic field describes a helical orbit around the magnetic field lines, as shown in Figure 1–2. Consequently, the charged particle cannot move across the magnetic field lines and is thus confined as far as directions perpendicular to the magnetic field are concerned. The particle is free to move along the

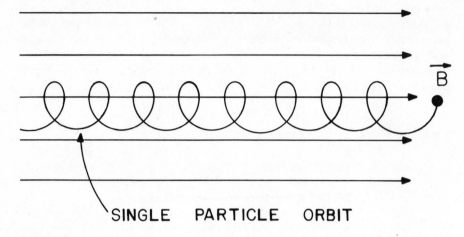

SINGLE   PARTICLE   ORBIT

Figure 1–2. Single Particle Orbit in a Uniform Magnetic Field.

magnetic field, however, and is thus not confined in this direction. Confinement of the particle along the magnetic field can be achieved in at least two ways. First, at each end of a region of constant uniform magnetic field, the field strength can be increased so that the magnetic field lines are squeezed closer together, as shown in Figure 1–3, to form what is called a magnetic bottle. From the Lorentz force $\mathbf{F} = q\mathbf{v} \times \mathbf{B}$ (q = charge on the particle, $\mathbf{v}$ = particle velocity) on the particle, it is easy to see that a particle moving into a region of increasing magnetic field experiences a decelerating force directed along the axis because of its orbital velocity and the radial magnetic field component, $\mathbf{B}_r$. Thus, if the particle does not have excessive kinetic energy directed along the magnetic field, it will be reflected from the high field region back into the region of uniform magnetic field. For this reason, the regions of increased magnetic field are called magnetic mirrors. Although the mirrors are not perfectly reflecting (since a particle can penetrate them if it has sufficient kinetic energy directed along the magnetic field), they can substantially improve the confinement along the magnetic field.

A second means of accomplishing confinement of a single charged particle along the magnetic field is to bend the magnetic field lines into circles, as shown in Figure 1–4a. In this configuration there are no ends, so magnetic mirrors are not required. Confinement in this toroidal configuration is not perfect, however. Bending the magnetic field lines into circles means that the magnetic field is stronger near the inside (r_1) than near the outside (r_2) (see Appendix 1–A). This gradient in the magnetic field causes the particle to drift vertically (perpendicular to $\mathbf{B}$) across the magnetic field lines, and eventually to be lost as shown in Figure 1–4b.[j] This trans-

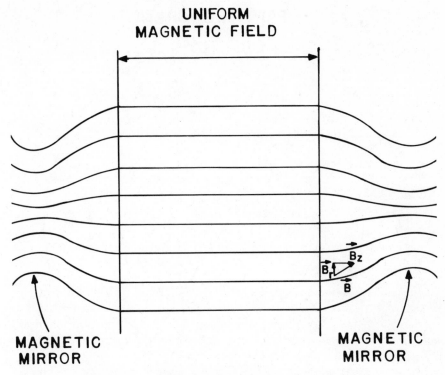

Figure 1–3. Magnetic Mirror Field Configuration.

verse drift can be made to average to zero for most particles by making the magnetic field lines spiral around the torus to form toroidal helices. Figure 1–4c, a right section of the torus, shows the points of intersection of one of the helical field lines with the plane of the figure after one, two, three, . . ., eight trips around the torus. The angular displacement ι (iota) of the line after each trip around the torus is constant and is called the rotational transform. Unless $\iota = 2\pi/n$, n a rational number, the points of intersection fill in a closed curve.[k] Since a particle tends to follow magnetic lines of force, it tends to move along the curve C in Figure 1–4c as it travels around the torus and thus moves alternately toward the top and the bottom of the torus. Consequently, the transverse drift motion (vertical in Figure 1–4b) of the particle moves it alternately toward and away from the magnetic axis M (Figure 1–4c) near the center of the magnetic field region. If the particle moves around the torus quickly enough and the rotational transform ι is large enough, then the drifts toward and away from the magnetic axis cancel, on the average, and the particle is confined by the toroidal field. A single particle that moves too slowly along the magnetic field lines will be lost, however.

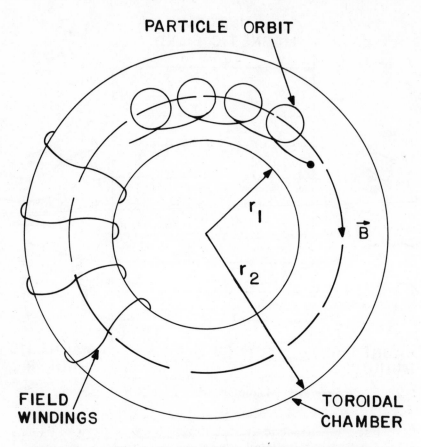

Figure 1–4a. Simple Toroidal Magnetic Field Configuration.

It is interesting to note that single particles with too much kinetic energy along the field lines are lost from magnetic mirror fields but those with too little kinetic energy along the field lines are lost from toroidal fields. Thus, although both magnetic mirror and toroidal magnetic field configurations can confine single particles effectively, the confinement is not perfect in either case.

Magnetic Confinement of a Plasma

Thus far, we have considered confining only a single particle in a magnetic field. When we think about confining a group of particles, the case of interest for nuclear fusion, additional complications arise which reduce

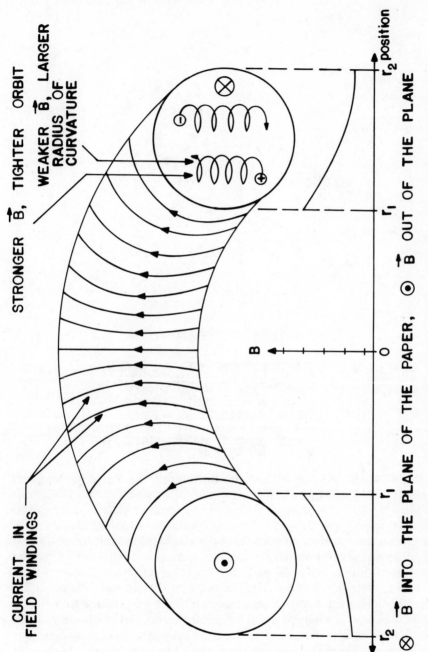

Figure 1–4b. Drift of Particles across a Simple Toroidal Confining Magnetic Field.

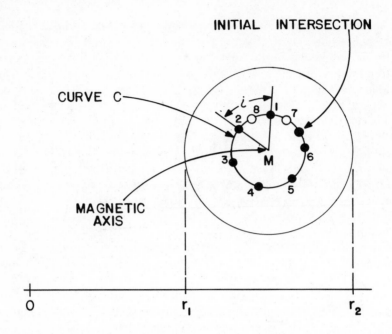

INITIAL INTERSECTION

CURVE C

MAGNETIC AXIS

ι ROTATIONAL TRANSFORM

●,○   INTERSECTION OF A GIVEN HELICAL FIELD LINE WITH THE PLANE OF A RIGHT SECTION OF THE TORUS. OPEN CIRCLES SIGNIFY INTERSECTIONS AFTER THE FIELD LINES HAVE MADE A 2π ROTATION.

Figure 1–4c. Rotational Transform Introduced by Spiraling Magnetic Field Lines.

the confinement compared with that for a single particle. For one thing, if we try to put several positive nuclei into a magnetic bottle, the Coulomb repulsion works to push the particles out. This repulsion can be reduced by adding electrons to neutralize the positive charges on the nuclei. The effect is minimized when enough electrons have been added to make the total collection of charged particles neutral. This circumstance is fortunate since it means that we can create the proper mixture of nuclei (ions) and electrons simply by ionizing the particles in a neutral gas. Such a collection of electrons and ions forms a plasma.

A second complication results from collisions between the particles in the plasma. Collisions enable the particles to move across the magnetic field lines to the edge of the field region and be lost. The collisions thus lead to diffusion of the charged particles across the magnetic field lines. Consider, for instance, a particle orbiting about point a with an orbit radius R which collides with a heavy particle at point b as shown in Figure 1–5. The particle rebounds from the collision and begins to orbit about point c with some new orbit radius R'. Thus, because of the collision, the particle has moved a distance $R + R'$ across the magnetic field. Simple calculations show that the number of particles per unit area per second lost across the confining magnetic field lines due to collisions should be proportional to $1/B^2$. Thus, as the confining magnetic field is increased, the particle losses caused by collisions should decrease fairly rapidly. This result is, in part, due to the fact that as B is increased, the particle orbits become smaller, so that the distance a particle moves across the magnetic field as the result of a single collision is correspondingly reduced (see Figure 1–5).

A more serious complication that arises when we try to confine a group of particles in a magnetic field is particle losses due to oscillations of the plasma particles, or plasma instabilities, as they are called. The energy necessary to drive some instabilities comes from forces applied to the plasma to confine it.[1] The lowest energy state for a plasma of a given temperature occurs when the plasma is uniformly distributed through space. Clearly, a plasma of finite size, confined by external forces, is not in this lowest energy state, and thus has some additional internal energy. Generally, the plasma tries to reach the lower energy states by using the extra internal energy to drive instabilities that will let the plasma move across the magnetic field (as discussed below) and spread out more uniformly. In a sense, a confined plasma wants to spread out (become more uniformly distributed) in much the same way gas contained in a bottle wants to burst the bottle and spread out (diffuse). Instabilities provide a mechanism for the plasma to "punch holes" in the magnetic bottle and escape.

As with all oscillations, two elements are required for plasma instabilities. First, some kind of restoring force must be available. In mechanical oscillators, the restoring force is provided by a spring. In a plasma, the restoring force can be provided either by the confining magnetic field or by Coulomb forces between the charged particles. Second, some kind of inertia must be present. In a mechanical oscillator, the inertia is provided by the mass at the end of the spring. In a plasma, the inertia of the ions and electrons is involved.

Although there are very many different plasma instabilities, most of them can be classified either as MHD (magnetohydrodynamic) instabil-

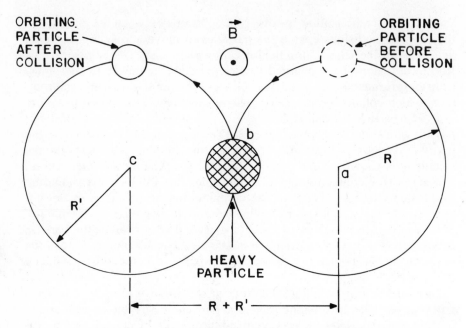

Figure 1–5. Effect of a Particle Collision.

ities or as microinstabilities. Roughly speaking, the MHD instabilities cause the ions and electrons to move together and the plasma behaves as a conducting fluid. These instabilities, also known as macroinstabilities, result in gross motion of the plasma and can cause the plasma to bump into the walls of the vacuum vessel in which the plasma is created.[m] Contact with the wall cools the plasma and thereby destroys its thermonuclear characteristics. (In a full scale reactor, the plasma would also burn a hole in the wall.) The MHD instabilities can be eliminated by designing the confining magnetic field so that it has a minimum at some point in space (where the plasma is located) and increases as one moves away from this point in any direction. Such a confining magnetic field is known as a minimum-*B* configuration. Since a plasma is generally diamagnetic[n] and diamagnetic materials are attracted to regions of low magnetic field[o] the plasma is in a (magnetic) potential well and a state of stable equilibrium. A plasma placed in a minimum-*B* confining field is therefore stable against gross fluid (MHD) motion.

Microinstabilities,[p] as we mentioned, involve relative motion between ions and electrons. As the ions and electrons separate, local electric fields are generated. If the amplitude of these instabilities becomes large enough, nonlinear effects produce a kind of turbulence and with it electron-ion separations large enough to generate very high local electric fields. These localized regions of very high electric fields give a charged particle a "kick" very much like it experiences when it collides with another charged particle. In fact, to one charged particle, another charged particle looks much like a localized region of high electric field. Thus, it is not surprising that the regions of high electric field produced by the microinstabilities can bump charged particles across the confining magnetic field and thus lead to enhanced particle loss. This enhanced particle loss due to microinstabilities is often called Bohm or anomalous diffusion. Empirically, it has been found that the number of particles per unit area per second lost across the confining magnetic field lines due to Bohm diffusion goes as $1/B$. Recall that the corresponding classical particle loss due to collisions goes as $1/B^2$. Thus, Bohm diffusion is much more difficult to reduce by increasing the strength of the confining field than the particle diffusion due to collisions. In fact, it turns out that Bohm diffusion cannot be slowed enough by increasing B to make power production from thermonuclear fusion economically feasible. The only alternative is to try to prevent the microinstabilities that cause Bohm diffusion. Theory developed largely during the decade of the 1960s indicates that many of these instabilities can be prevented by introducing shear into the confining magnetic field. By shear, we mean that the pitch of the helical field lines (see discussion of Figure 1–4) varies as we move radially out from the center of the plasma column. Thus, the "direction" of the magnetic field changes as we move in any direction away from a point in the plasma. Most instabilities have a preferred orientation with respect to the magnetic field lines. When the confining magnetic field has shear, the field lines are no longer parallel and extend in no single direction. Thus, the instabilities cannot properly orient themselves with respect to a sheared magnetic field and are consequently suppressed.[q] It follows, however, that microinstabilities that have wavelengths short compared with the distance over which appreciable shear occurs are not suppressed. Thus, the short wavelength (and hence high frequency) microinstabilities are the most difficult to prevent. It is thus important to find plasma-magnetic configurations that minimize the energy available to drive those instabilities so that even if they are not stabilized, their amplitudes are limited to small values.[r]

We note in passing that some plasma instabilities have also been suppressed by using dynamic stabilization and negative feedback. Dynamic stabilization is based on the fact that a system in a state of unstable equilibrium can, under certain conditions, be made stable by applying a suitable oscillatory force to jiggle the system. For example, an inverted pendulum can be made to stand upright by jiggling its pivot vertically at a suitable frequency. With negative feedback, on the other hand, some signal related to the instability is measured, amplified and inverted, and then applied as a corrective signal (perhaps as a corrective magnetic field) to the plasma. It is not yet clear that either of these techniques will be important in stabilizing a reactor plasma because of their relative complexity and because of the difficulty in designing them to deal with more than one mode of instability. Since there are many modes of instability in a plasma, stabilizing only one may simply permit another to grow.

For many years, the particle loss rate in most thermonuclear fusion experiments was dominated by Bohm diffusion. Experiments beginning in the late 1960s indicate that particle losses due to Bohm diffusion can actually be drastically reduced, if not eliminated, for some kinds of plasmas by careful design of the confining magnetic field to include shear and the like. During the 1970s, more experiments are being performed to see whether or not Bohm diffusion can be kept small in plasmas of the kind needed for thermonuclear fusion.

Radiation Losses from a Plasma

We have seen how collisions and instabilities can lead to particle losses from the plasma. These particle losses reduce the thermonuclear power produced in the plasma by letting particles escape from the plasma before fusing. In contrast, the radiation losses discussed in this section reduce the thermonuclear power released by cooling the plasma so that fewer particles have sufficient energy to overcome the Coulomb barrier and fuse [6]. It is well known from classical electromagnetic field theory that a charged particle ordinarily gives off radiation whenever it is accelerated. Generally, the more it is accelerated, the more it radiates. In a plasma, the ions and electrons are accelerated by at least two mechanisms. First, the forces exerted on a charged particle as it encounters other charged particles in the plasma produce acceleration. Radiation produced by this mechanism is called bremsstrahlung, where "brems" means braking or decelerating in German and "strahlung" means radiation. Second, the ions and electrons feel a radial force that causes them to orbit around the magnetic field lines and this force gives rise to acceleration. Radiation produced by this mechanism is called synchrotron radiation.

It is easy to see that the electrons in a hydrogen plasma do most of the radiating, while the ions do very little. The Lorentz force on the ions and electrons is $\mathbf{F} = q(\mathbf{E} + \mathbf{v} \times \mathbf{B})$. The magnitude of the acceleration is, from Newton's second law, $a = |\mathbf{F}|/m$, where m is the mass of the particle. Since m is very much smaller and $|\mathbf{v}|$ is typically much larger for the electrons than for the ions, the electrons clearly are accelerated much more and, therefore, are the dominant source of radiation. We therefore neglect the radiation due to the ions as compared to that of the electrons.

The largest radiation loss from a thermonuclear fusion plasma is likely to be bremsstrahlung. The bremsstrahlung power radiated per unit volume is given by

$$P_{br} = 1.5 \times 10^{-38} Z^2 n_i\, n_e T_e^{1/2} \quad \text{MW/m}^3 \tag{1.3}$$

where Z is the atomic number of the ions

n_i is the number of ions per m³

n_e is the number of electrons per m³

T_e is the electron temperature in eV

Notice that the bremsstrahlung power density goes as Z^2. Thus, another factor appears which favors low Z elements for nuclear fusion fuel.[s]

Typical thermonuclear fusion plasmas emit most of the bremsstrahlung in the soft x-ray portion of the electromagnetic spectrum. Since the plasma is almost transparent to such radiation, most of the bremsstrahlung energy will pass through the plasma and must be absorbed and collected externally, if it is not to be wasted.

The power in synchrotron radiation can also be appreciable. For a hydrogen isotope plasma, the power radiated per unit volume is given approximately by

$$P_{\text{sync}} \approx 10^{-19} n_e B^2 T_e \quad \text{W/m}^3 \qquad n_e \text{ in (m}^{-3}) \tag{1.4}$$

$$B \text{ in (tesla)}$$

$$T_e \text{ in (eV)}$$

where B is the confining magnetic flux density. Note that synchrotron radiation increases much more rapidly with temperature than does bremsstrahlung. At thermonuclear plasma densities and temperatures, synchrotron radiation can even be larger than bremsstrahlung. Fortunately, the synchrotron radiation is emitted in the infrared and microwave portions of the electromagnetic spectrum. In this region, the plasma is somewhat opaque so that it absorbs some of the radiation. In addition, reflectors could be placed at the wall of the vacuum vessel to reflect the radiation through the plasma several times and insure almost complete absorption of the synchrotron radiation. For this reason, radiation loss from a plasma by synchrotron radiation is likely to be less important than the bremsstrahlung loss as far as cooling the plasma is concerned.[t] From this point on, therefore, we neglect synchrotron radiation losses in comparison to bremsstrahlung losses.

Notes

(a) An amu is very nearly the mass of either a proton or a neutron and is precisely defined to be 1/12 of the mass of a $_6C^{12}$ atom. (In this notation, the subscript is the atomic number of the element while the superscript is the mass number of the nucleus.) With this particular choice for the size of 1 amu, the mass of any nucleus in amu, rounded off to the nearest integer, is just the mass number (the total number of protons and neutrons in the nucleus). This fortunate circumstance would not occur if, for example, we defined 1 amu to be exactly the mass of the proton or neutron.

(b) This relation assumes that the volume of the nucleus V is proportional to its mass number A and that the nucleus is roughly spherical so that $V \propto R^3$. Thus, $R \propto A^{1/3}$. The constant of proportionality is empirical.

(c) $1 \text{ eV} = 1.60 \times 10^{-19}$ Joule.

(d) In addition to these direct fusion reactions certain catalytic (chain) fusion reactions may also be of interest. See, for example, J. R. McNally, Jr., *Nuclear Fusion* 11 (1971) 187, 189, 191, 554.

(e) The effective height of the Coulomb barrier is, in fact, reduced by the tunnel effect. This is a quantum mechanical process in which a particle that lacks sufficient energy to go over the top of a Coulomb potential barrier can nevertheless penetrate or tunnel through it, provided the barrier is sufficiently thin. The tunnel effect for electrons is fundamental in the operation of the tunnel diode, for example.

(f) A possible exception is the somewhat controversial Migma device which is based on the colliding beam storage concept from high energy nuclear physics (see B. C. Maglich, J. P. Blewett, A. P. Colleraine, and W. C. Harrsion, "Fusion Reactions in Self-Colliding Orbits," *Phys. Rev. Lett.* 27, 909(1971). Also, see F. F. Chen et al. *Princeton Plasma Physics Laboratory Report MATT-1237*, April 1976).

(g) A similar scheme, in which a solid pellet containing deuterium would be bombarded by a beam of deuterium nuclei, fails because most of the deuterons simply ionize atoms in the pellet, while practically none fuse with the deuterium in the solid.

(h) For simplicity, we assume the particles have a Maxwell-Boltzmann velocity distribution so that the temperature is well-defined.

(i) A sort of hybrid combination of colliding beam and ordinary thermonuclear fusion processes has been proposed by John Dawson, Harold Furth, and F. H. Tenney (*Phy. Rev. Letters* 26, 1156 (1971)). In this approach, an energetic beam of ions is fired through a more or less stationary collection of target ions. As the ions in the beam collide with the target ions, both fusion and scattering processes occur. In contrast to the colliding beam approach, however, the scattered ions are not lost after a collision or two, but have many chances to collide and fuse, without being lost, just as in ordinary thermonuclear fusion. In contrast to simple thermonuclear fusion, however, it is not necessary to heat all the ions to a very high temperature so that a few of them will have enough energy to overcome the Coulomb barrier and fuse. Instead, as in the colliding beam approach, the energy is concentrated in the energetic ion beam whose ions have a much higher probability of overcoming the Coulomb barrier than the multitude of slower moving ions present in an ordinary thermonuclear reactor. The two ion component approach is therefore designed to realize the best features of both colliding beam and thermonuclear fusion. We will have occasion to discuss the two ion component approach in more detail later on.

(j) The mechanism for this "grad B" drift is easily understood on recalling that, in a magnetic field, the radius of curvature of the orbit of a charged particle varies inversely with B. As a charged particle orbits around the field lines of a spatially inhomogeneous magnetic field, therefore, the radius of curvature is small in regions of stronger magnetic field and larger in regions of weaker magnetic field. In the simple toroidal magnetic field of Figure 1–4b, the alternate increase and decrease in its orbit radius of curvature as the particle passes alternately through regions of weaker and stronger magnetic fields, causes the particle to drift vertically. Positive and negative particles drift in opposite directions because they orbit in opposite directions in a magnetic field.

(k) Note that the field lines trace out what is called a "magnetic surface" as they encircle the torus.

(l) Other instabilities (called velocity space instabilities) are driven by the energy available when the velocity distribution of a plasma is not Maxwell-Boltzmann, the velocity distribution a plasma ordinarily prefers. These deviations from a Maxwell-Boltzmann velocity distribution correspond to potential or free energy and may occur, for example, when certain methods are used to heat the plasma or when particles with velocities in certain ranges escape more readily than others. In magnetic mirror devices, the escape of particles with large axial velocities produces a non-Maxwell-Boltzmann velocity distribution. The resulting instabilities are called "loss cone" instabilities.

(m) As an example of a mechanism that can cause the plasma to move as a conducting fluid recall the grad B drift for single particles described in a previous note. In the inhomogeneous magnetic field of a torus (Figure 1–4b), the positive and negative particles drift apart and hence produce a vertical electric field, $\mathbf{E}$. This vertical velocity in conjunction with the confining magnetic field, $\mathbf{B}$, produces a Lorentz force $\mathbf{F} = q\mathbf{v} \times \mathbf{B}$ on the particles which pushes them in the direction of the vector $\mathbf{E} \times \mathbf{B}$ regardless of the sign of this charge. (For a more detailed discussion of the crossed field or $\mathbf{E} \times \mathbf{B}$ drift, see, for example, F. F. Chen, *Introduction to*

Plasma Physics, Plenum Press, New York, 1974, p. 23). Thus, the plasma as a whole drifts to the outside wall of the torus. This particular MHD plasma drift is at least partially suppressed in a toroidal field with a rotational transform (Figure 1–4c) since the electrons can move along the helical field lines, from bottom to top, and hence short out the electric field induced by the magnetic field gradient. Note, however, that if the rotational transform ι is some multiple of 2π, then the field line closes on itself and hence does not trace out a magnetic surface as it encircles the torus. The access of the electrons to various regions of the plasma by moving along the magnetic field lines is thus drastically curtailed. They are therefore no longer free to short out the electric field and hence suppress the outward drift of the plasma. For an equilibrium configuration of the plasma and the magnetic field to exist, therefore, we must require $\iota \neq 2\pi$ so that the field lines do not close on themselves. Furthermore, analysis shows that if the equilibrium is to be stable, then we must require $\iota < 2\pi$. This condition is the so-called Kruskal-Shafranov limit. Physically, it results from the fact that if the magnetic field lines spiral too quickly around the torus, the plasma can easily begin to kink and hence go unstable. Such "kink" instabilities, as they are called, are one type of MHD or macroscopic instability.

(n) That a plasma is usually diamagnetic is indicated by the fact that both the ions and electrons in a plasma in a steady magnetic field rotate (in opposite directions) so that each particle in effect produces a tiny loop current which in turn generates a magnetic field in a direction opposite to the applied magnetic field. Since the net magnetic field in the region is therefore reduced by the presence of the plasma, the plasma is diamagnetic.

(o) This effect is opposite to that observed in paramagnetic and ferromagnetic materials, which are attracted to regions of high magnetic field. An iron nail, a ferromagnetic body, is attracted to the surface of a magnet, the region of highest field. Within paramagnetic and ferromagnetic materials, the magnetic field is larger than the applied magnetic field.

(p) As the name suggests, no gross plasma motion occurs with microinstabilities.

(q) Shear can similarly be used to help suppress some MHD instabilities.

(r) Because of their relatively high frequencies, microinstabilities fortunately tend to saturate at relatively low amplitudes, compared with macroinstabilities, because of nonlinear effects. Specifically, if the microinstability has frequency ω and particle excursion amplitude A, then the velocity of the particles involved in the instability is something like ωA so that the kinetic energy of the particles (due to the instability) is proportioned to $(\omega A)^2$. For a limited amount of energy available to drive instabilities, therefore, a high frequency wave tends to be limited to lower amplitudes. Hopefully, therefore, the short wavelength, high frequency microinstabilities which are the hardest to stabilize may not be too troublesome.

(s) Notice that impurities with high Z can drastically increase the bremsstrahlung loss. Fusion plasmas must therefore be essentially free of impurities, except for certain controlled amounts of impurities which might be introduced for temperature control.

(t) Under certain conditions, synchrotron radiation may dominate despite the use of reflecting walls (see M. N. Rosenbluth, *Nuclear Fusion* 10, 340 (1970)). If this circumstance should occur, some of the following results (ideal break-even temperature, Lawson criterion) would be modified.

Appendix 1–A

Relation of Magnetic Field Curvature and Magnetic Field Gradients

Consider a torus in which a steady current flowing in a coil wound on the torus produces a toroidal magnetic field **B** as shown in Figure 1–A–1. If the current is characterized by a current density **J** then $\nabla \times \mathbf{B} = \mathbf{J}$. If we integrate this equation over a circular disc (1) which is centered on the major axis, (2) which lies in a plane parallel to the midplane of the torus, and (3) whose perimeter lies within the torus, we find that

$$\int (\nabla \times \mathbf{B}) \cdot d\mathbf{A} = \int \mathbf{J} \cdot d\mathbf{A} = \text{constant}$$

Using Stokes's theorem and symmetry,

$$\int \nabla \times \mathbf{B} \cdot d\mathbf{A} = \oint \mathbf{B} \cdot d\mathbf{l} = 2\pi r B$$

Thus

$$2\pi r B = \text{constant}$$

$$B = \frac{\text{constant}}{r}$$

This result clearly shows that the curved magnetic field inside the torus has a spatial gradient. In fact, it is a general property of curved magnetic fields to have spatial gradients. As an example, consider a magnetic field in a volume in which there is no current flowing (although current may be flowing outside the volume). Then $\nabla \times \mathbf{B} = \mathbf{J}$ becomes simply $\nabla \times \mathbf{B} = 0$. Suppose that this magnetic field in the volume has a curvature which can be simply described in cylindrical coordinates by $B_r = 0$, $B_\theta \neq 0$, $B_z = 0$. Then, $\nabla \times \mathbf{B} = 0$ becomes $\partial B_\theta / \partial z = 0$ and

$$\frac{1}{r} \frac{\partial}{\partial r}(r B_\theta) = 0$$

Since $\nabla \cdot \mathbf{B} = 0$ (in the case $B_r = 0$ and $B_z = 0$) tells us that $\partial B_\theta / \partial \theta = 0$, we see that B_θ depends only on r. Thus our equations become

$$\frac{d}{dr}(r B_\theta(r)) = 0, \qquad r \neq 0$$

or

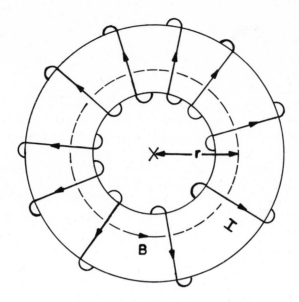

Figure 1–A–1. Inhomogeneous Magnetic Field in a Torus.

$$B_\theta(r) = \frac{\text{constant}}{r}$$

This example illustrates the general result that magnetic field curvature results in a field gradient such that the magnetic field decreases away from the center of curvature. The converse (that field gradients produce field curvature) is not true in all cases.

2 Controlled Fusion Reactors

Plasma Parameters for a Thermonuclear Reactor

We now estimate the ranges of density, temperature, and magnetic field for plasmas in thermonuclear reactors [6].

Power Density and Particle Density in a
Thermonuclear Reactor

Power densities for D-T and D-D (both branches, as shown in equations (1.2)) thermonuclear reactions as a function of deuteron density are shown in Figure 2–1 for temperatures of 10 keV and 100 keV. These temperatures are somewhat lower than the several hundred keV mentioned earlier as the approximate energy needed to overcome the Coulomb barrier because it is unnecessary for all the ions to overcome the barrier before large amounts of fusion energy can be released. (This is fortunate because higher temperatures increase radiation losses and require higher, more expensive magnetic fields for confinement.) Notice, though, that the curves in Figure 2–1 are not very sensitive to temperature. Thus, we can determine a suitable range of densities more or less independent of temperature by considering the thermonuclear power density produced in the plasma.

If the power density is chosen too low, a large volume of plasma, and hence a spatially large magnetic field, would be required to produce a reasonable output power. The capital cost of the power plant would then be excessive. For example, a deuteron density of 10^{18} m^{-3} produces less than 1 kW/m^3. Since a typical power plant produces more than 10^8 W, a plasma volume of 10^5 m^3 would be required. This volume corresponds to a linear dimension of about fifty meters. The cost of producing a magnetic field strong enough to confine even a 10-keV plasma with a density of 10^{18} m^{-3} and a volume of 10^5 m^3 is enough in itself to rule out any possibility of economic power production. If, on the other hand, the power density is chosen too high, materials problems caused by the intense neutron flux in the reactor are impossible to solve. For example, a deuteron density of 10^{25} m^{-3} produces 10^{15} W/m^3. Thus, only 10^{-7} m^3 (0.1 cm^3) of a plasma with 10^{25} m^{-3} deuteron density would produce as much power as a typical

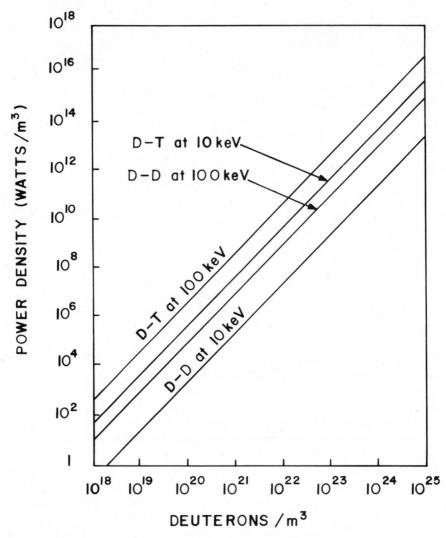

Figure 2-1. Thermonuclear Power Density versus Particle Density for D-T and D-D Reactions.

conventional steam plant. Such power density obviously would lead to impossible materials problems.

In nuclear fission reactors, the design value of the power density in the core is typically 20-60 MW/m³. Although this value is chosen by considering materials problems somewhat different from those in a nuclear fusion reactor, 40 MW/m³ is not likely to be greatly different from the value a careful design of a fusion reactor would give.[a] For one thing, the volume required to produce 10^9 W is 25 m³, corresponding to a linear dimension of about 3 m. Magnetically confining a 10-keV plasma of this volume is not out of the question economically. Referring to Figure 2–1, we therefore expect the plasma density in a fusion reactor to be in the range 10^{20}–10^{21} m⁻³ if the power density is about 40 MW/m³. These densities are only small fractions (10^{-5}) of the density of a gas at standard temperature (27°C) and pressure (1 atmosphere). The plasma region of a fusion reactor is thus essentially a vacuum.

The Ideal Break-Even Temperature

The minimum operating temperature for a self-sustaining thermonuclear reactor is that at which the energy released by nuclear fusion just exceeds that lost from the plasma as a result of radiation losses, primarily bremsstrahlung. This temperature, which we call the ideal break-even temperature [6], is ideal in that confinement is assumed to be perfect so that no energy is lost by means of escaping particles.[b] Figure 2–2 shows fusion power density released and bremsstrahlung loss as functions of temperature for both D-D and D-T reactions for an ion density of 10^{21} m⁻³.

From this figure, we see that the ideal break-even temperatures for the D-T and D-D reactions, are about 4 keV and 40 keV respectively. The relatively low ideal break-even temperature of the D-T reaction is a very desirable property for at least two reasons. First, the lower temperature means that it should be easier to heat a D-T plasma to a point at which significant energy is released by fusion. Second, the lower temperature means lower plasma kinetic pressure ($n_e kT_e + n_i kT_i = 2nkT$ in case the electron and ion densities and temperatures are about the same) and hence lower (cheaper) magnetic fields to confine it. Notice that a D-T reactor with an ion density of 10^{21} m⁻³ should be operated at temperatures of about 20 keV to achieve a power density in the range of 40 MW/m³.

The Lawson Criterion

We have just seen that if plasma confinement were perfect in a reactor so that radiation were the only means of energy loss, then we could achieve energy break even in the reactor by simply heating the plasma to some critical temperature. We now consider the more realistic case in which the plasma is confined only for some finite time τ so that energy is lost to escaping particles as well as to radiation. We expect, of course, the temperature required for the reactor to be self-sustaining to increase as the confinement gets worse.

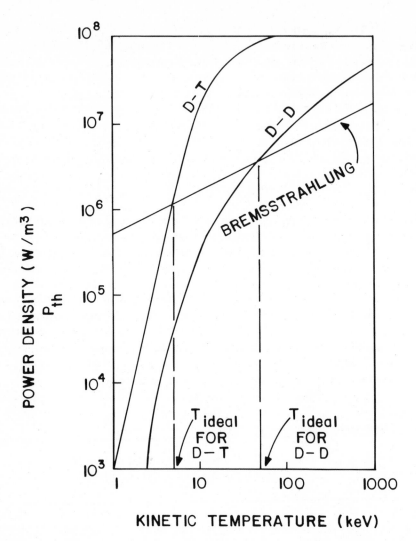

Figure 2–2. Ideal Break-Even Conditions for D-D and D-T Fusion Reactions.

Consider a pulsed reactor in which the fuel is heated rapidly to a temperature T and then confined and maintained at this temperature for a time τ while thermonuclear reactions occur. Then, suppose the plasma is cooled and the reactor is refueled before the next pulse. If the gas density of the neutral hydrogen fuel atoms is n, then the charged particle density of the completely ionized hydrogen isotope plasma is $2n$ (half electrons, half ions). Assuming that both the ions and the electrons have a temperature T, the average energy per particle is $3/2\ kT$. Thus, the energy density required to heat the fuel to a temperature T is about $3\ nkT$ since the initial gas temperature is very small compared with the temperature T of the fusion plasma, and the ionization potential of the hydrogen atoms is small compared to $3/2\ nkT$.

Let P_{th} be the thermonuclear fusion power density and P_{br} be the bremsstrahlung power density in the plasma. If the bremsstrahlung is collected and absorbed outside the plasma, as suggested earlier, then the total energy available after the thermonuclear reaction pulse in a plasma of volume V is

$$E_{\text{out}} = (\tau P_{th} + \tau P_{br} + 3nkT)V \qquad (2.1)$$

since by hypothesis we maintain the temperature at T throughout the pulse. Suppose the overall efficiency for using this energy to heat the plasma is $1/3$. Then the minimum condition for self-sustaining operation of the reactor [6] is that the available output energy, $1/3\ (\tau P_{th} + \tau P_{br} + 3nkT)V$ be greater than the input energy,[c] $(3nkT + \tau P_{br})V$. The input energy to the plasma follows from observing that to create the plasma with temperature T we must supply an energy of $(3nkT)V$, and to maintain the plasma at temperature T we must replace the energy $(\tau P_{br})V$ lost from the plasma by bremsstrahlung. Thus, to get out more energy than it takes to run the reactor, we require that

$$\frac{1}{3}(\tau P_{th} + \tau P_{br} + 3nkT)V > (\tau P_{br} + 3nkT)V \qquad (2.2)$$

Rearranging this inequality, we get

$$\frac{P_{th}/(3n^2kT)}{P_{br}/(3n^2kT) + \dfrac{1}{n\tau}} > 2 \qquad (2.3)$$

We saw in equation (1.3) that P_{br} increases as n^2, and Figure 2–1 shows that P_{th} is also proportional to n^2. Thus, the left hand side of the inequality really depends only on the quantities T and $n\tau$. For D-T, the minimum value of $n\tau$ that satisfies the inequality is[d] $n\tau \approx 10^{20}$ m^{-3}-sec, which occurs for $T \approx 13$ keV. The smallest value for $n\tau$ that satisfies the inequality is relatively independent of temperature, however, for 10 keV $< T < 15$ keV. For D-D, the minimum $n\tau$ is $\sim 10^{22}$ m^{-3}-sec at a temperature $T \approx 100$ keV. These minimum values depend, of course, upon the efficiency with which the available energy can be converted into useful power. If this efficiency exceeds $1/3$ (the value used above), the minimum values of $n\tau$ will be somewhat less than those given above. However, more sophisticated calculations, allowing for nonuniform plasmas and so forth actually lead to even higher value of $n\tau$.[e]

A striking feature of these results is the substantially lower (two orders of magnitude) value of $n\tau$ required for D-T than D-D reactions. For a given plasma density, this means that a pulsed D-T reactor needs to be operated only one percent as long as a D-D reactor to reach the break-even point in energy. Since, during a long pulse, plasma can move across the magnetic field lines (by means of collisions or instabilities) and be lost, the D-T reaction with its shorter required pulse length is clearly preferable in this regard. The D-T reaction also has a lower ignition temperature, another attractive feature. The D-T reaction does, however, have two major disadvantages: (1) tritium is radioactive (12.4 years half-life) and therefore does not occur in nature (it must be generated), and (2) about eighty percent of the energy from this reaction comes off in neutrons. As we will discuss later, neutron kinetic energy may not be the worst way in which the energy could appear, but it is not the best by any means. Both of these factors make D-T thermonuclear fusion reactor design extremely difficult from an engineering viewpoint.

Although this discussion has considered only pulsed reactors, similar results must also apply to a reactor run in steady state. In this case, τ would represent some average confinement time of the reacting nuclei instead of the pulse length.

Magnetic Field Strength

We have seen how a magnetic field confines a plasma by retarding the motion of the charged particles across the magnetic field. If we envision this confining force to be produced by an effective magnetic pressure, then its size turns out to be $(B_0^2 - B^2)/2\mu_0$, where B_0 and B are the magnetic flux densities outside and inside the plasma, respectively, and μ_0 is the magnetic permeability of free space. For successful confinement of the plasma, the magnetic (confining) pressure must be at least as large as the kinetic pressure, $2nkT$, produced by the plasma.[f] If no magnetic field penetrates the plasma so that there is no magnetic field inside it, then the magnetic pressure is $B_0^2/2\mu_0$. In this case ($\beta = 1$), we can solve the minimum field required for confinement:

$$B_{\min} = \sqrt{4\,\mu_0 nkT} \tag{2.4}$$

We decided from energy density considerations that $10^{20}\text{m}^{-3} < n < 10^{21}\text{m}^{-3}$ for the D-T reaction, and found that the Lawson break-even temperature $T_{\text{ideal}} \approx 10$ keV. With $T \approx 10$ keV, we obtain a power density of about 40 MW/m^3 with $n \approx 5 \times 10^{20}\text{m}^{-3}$. Then $B_{\min} \approx 2.0$ T. In practice, some of the magnetic field penetrates the plasma after a short time since the plasma is not perfectly conducting ($\beta < 1$).[g] Thus, $B \neq 0$ and hence we must have $B_0 > B_{min}$. More detailed calculations typically require B_0 to be at least three or four times B_{min} to achieve adequate confinement. That is $B_0 \approx 7\,T$. This value is fortunately well below $10\,T$, sometimes mentioned as the technological limit for magnetic fields produced in large volumes by superconducting magnets.

For the D-D reactions, T_{ideal} is about ten times as large as for the D-T reaction. Thus, we would expect B_{min} to increase by a factor $\sqrt{10} \approx 3.2$ in this case. If we suppose that B increases by a similar factor, we find that $B \approx 22\,T$ for the D-D reaction. The large value of the magnetic field required is an important disadvantage of D-D reactions in comparison with D-T reactions.

Summary of Plasma Parameters

A summary of the results of this section is given in Table 2–1. Notice that confinement times τ in the range 0.1–1.0 sec are required. Experimentally achieving a plasma with the $n\tau$ value of 10^{20} m^{-3}-sec and the break-even temperature of 10 keV (values determined for the D-T reaction) is the much talked about "scientific feasibility" criterion which the major plasma research laboratories in the world are striving to reach. We mention in passing that the best $n\tau$ values in today's (1976) experiments[h] are near 10^{19} m^{-3}-sec at a temperature $T \approx 1$ keV.

Design Considerations for a D-T Thermonuclear Fusion Reactor

The D-T reaction will probably be used in the first fusion reactor because of its low ignition temperature and low minimum value of $n\tau$. Unfortunately, the D-T reaction has some inconvenient features when it comes to designing a reactor to convert the fusion energy into useful power. In particular, a D-T reactor must convert the kinetic energy of the neutrons into useful energy and generate tritium to burn as fuel in the reactor. Of course, any type of fusion reactor must heat the fuel to fusion temperature and confine the plasma long enough to recover an economically interesting amount of energy. Accomplishing these tasks in a working reactor presents formidable engineering problems.

Blanket

Most of the energy (14.1 MeV out of 17.6 MeV) from a D-T reaction appears as kinetic energy of the neutrons. Neutron kinetic energy is not the

Table 2–1
Summary of the Requirements for Controlled Fusion

D–T	*D–D*	*Why necessary?*
10^{20} m^{-3} $\lesssim n \lesssim 10^{21}$ m^{-3}	10^{20} m^{-3} $\lesssim n \lesssim 10^{21}$ m^{-3}	Obtain suitable power density
$n\tau \gtrsim 10^{20}$ m^{-3}-sec	$n\tau \gtrsim 10^{22}$ m^{-3}–sec	Reach energy break-even point
$T > 10$ keV	$T > 100$ keV	
$B \gtrsim 7$ T	$B \gtrsim 22$ T	Achieve adequate confinement time

most convenient and flexible form of energy, however. In fact, about the only thing we can do with neutron kinetic energy is to convert it into heat and then use a thermal cycle to convert the heat to electrical energy.[i] But using a thermal cycle means that the conversion efficiency is limited by the Carnot efficiency, and in practice to an absolute maximum of about sixty percent (forty percent is typical in a present day power plant) by the lack of suitable high temperature materials. If most of the energy from the fusion reaction came off in charged particles instead of neutral particles, we could probably use some direct energy conversion scheme and produce electrical power at efficiencies higher than sixty percent. This is one reason for the interest in the D-He^3 reaction in which all the reaction products are charged so that direct energy conversion should be possible.[j]

The high energy neutrons cause additional problems by irradiating the structural materials of the reactor. The irradiation makes these materials radioactive and may cause structural damage to them as well. The effects of irradiating materials with a high flux of 14-MeV neutrons are actually not known, but, based on experience with lower energy neutrons in fission reactors, they are not expected to be good. There are presently no 14-MeV neutron sources with high enough flux ($> 10^{13}$ cm^{-2} sec^{-1}) available to carry out testing of materials under reactor conditions.[k] The consequent lack of information about how materials will hold up in a reactor is perhaps the major difficulty in designing a D-T reactor. With reactions other than D-T, the neutron irradiation problem is greatly reduced.

Generating tritium in the reactor for fuel is necessary because tritium, as stated previously, has a half life of only 12.4 years. Tritium generation can be accomplished by slowing down (moderating) the 14-MeV neutrons coming out of the plasma and then bombarding Li^6 in the blanket with them to obtain the following reaction,[l]

$$\text{Li}^6 + \text{n} \rightarrow \text{He}^4 + \text{T} + 4.6 \text{ MeV} \qquad (2.5)$$

Note that this reaction gives off additional energy and requires Li^6 as a fuel. Because the reactor both burns and generates tritium, one could argue that tritium is not a fuel at all, but merely a kind of catalyst for burning D and Li^6. In any case, we should note that it may not be a completely trivial matter to extract the T from the Li^6 and make it available to burn.[m]

Fuel Injection and Heating

There is some concern that new fuel injected into an operating reactor cannot reach the center of the plasma, but will be ionized near the edge of the extremely hot plasma and held there by the magnetic field. Advocates of pulsed (as opposed to steady state) reactors point out that this problem can be avoided in pulsed reactors simply by refueling between pulses when no plasma is present. We must note, however, that pulsed reactors are not without problems either. For example, these reactor schemes ordinarily use pulsed magnetic fields. These in turn are produced by pulsed currents flowing in conventional (as opposed to super-) conductors. Two problems arise. First, there may be appreciable ohmic losses of energy. This not only wastes energy, but also makes cooling the wall more difficult. Second, the energy stored in the coil during the pulse must be stored someplace else between pulses. Fast, low-loss energy storage is expensive and difficult to achieve. We cannot simply throw away the energy between pulses instead of storing it since the resulting power loss would be greater than the reactor output.

Heating the fuel to fusion temperature is clearly required regardless of the reaction used. In one way, the problem is less severe for the D-T reaction than for others because of its lower break-even temperature. Moreover, at least part of the fusion energy carried by the alpha particle (He^4 nucleus) is given up as heat to the plasma since the alpha particle is charged and hence confined by the magnetic field. The alpha particle initially moves at a high velocity, but is eventually slowed down, via collisions, as it moves through the plasma and gives up its energy. This energy added to the plasma can be used to heat the cold fuel being added and help compensate for radiation losses. In a continuous reactor, though, we may have to plan either on preheating the fuel before it is injected or on heating the plasma directly after the cold fuel has been added. If the reactor is operated so that alpha particle heating alone is sufficient to heat the new fuel and make up for radiation losses the reaction is said to be ignited. If supplemental heating is required, we speak of the power gain of the reactor as the ratio of its output power to the supplemental heating power required.[n]

With other reactions, such as D-He^3, all of the energy is carried by charged particles. This energy is deposited in the plasma since the charged particles (reaction products) are confined by the magnetic field. Heating of the fuel or plasma with auxiliary sources may therefore be unnecessary for these fuel cycles (ignition may be possible), but other problems related to instabilities may become more severe. (The release of high energy charged particles tends to distort the overall charged particle energy distribution function away from Maxwell-Boltzmann.)

In continuous reactors, plasma heating will probably be accomplished by ohmic heating together with neutral injection or rf heating. Ohmic heating consists of passing a large current through the plasma so that it is heated by I^2R dissipation. As the plasma temperature is increased by heating, however, its resistivity decreases to such an extent (to much less than that of copper) that ohmic heating becomes ineffective. Typical estimates of the maximum plasma temperatures that can be achieved with ohmic heating alone are in the few kilovolt range. Thus, if ohmic heating is used in a reactor, some supplemental means of heating must be used as well. Proposals have been made to increase the resistivity of the plasma at high temperatures by inducing turbulence in the plasma. As we discussed earlier, turbulence can increase the effective number of collisions a particle experiences and hence increase the plasma resistivity. Turbulence, then, might make heating a plasma by passing large currents through it effective at higher temperatures. Unfortunately, turbulence can also destroy confinement, as we discussed earlier. Experiments under way may show that certain kinds of turbulent heating do not lead to increased particle losses. Until such experiments are successful, however, the favored heating schemes for continuous reactors, in addition to ohmic heating, seem to be neutral injection and rf heating.

In neutral injection, a beam of high energy neutral deuterium and tritium is directed at the plasma. As the beam enters the plasma, the neutral particles are ionized and their motion is randomized by collisions with the plasma particles. It might seem, therefore, that neutral injection could be used to refuel the reactor and heat the plasma (with the neutral beam kinetic energy) simultaneously. In fact, however, the optimum (for penetration) injection energy for the neutral particles used for heating the plasma is high enough so that only a relatively tenuous beam would provide enough energy to heat the plasma. The upshot is that neutral beam heating usually will not provide enough hydrogen to fuel the reactor.[o]

With rf heating, certain radio frequency waves would be launched in the plasma and the wave energy absorbed by it as a means of heating. Compared with neutral injection, rf heating has the advantage of being based on currently available technology. It has the disadvantage of being difficult to couple into the plasma through the blanket if the intended wavelength is too long for waveguide transmission.

Compression of the plasma by increasing the magnitude of the confining magnetic field is a heating technique commonly proposed for pulsed reactors but which might also be useful in initially heating quasi-continuous reactors.[p] Other heating techniques mentioned for use with pulsed reactors include heating with shock waves launched in the plasma and heating with lasers or electron beams.

Regardless of the particular technique used, a primary consideration is that the heating process not drive plasma instabilities by, say, warping the velocity distribution function away from Maxwellian.

Ash Removal

The complementary problem to injecting new fuel into the reaction region is removing the reaction products or ashes from the same region. The ashes cannot be allowed to accumulate since they cause enhanced bremsstrahlung losses (because of their higher Z). In a pulsed reactor, separating out the reaction products can be accomplished between pulses by replacing the partially burned nuclear fuel with new fuel and removing the partially burned fuel for purification away from the reaction region. In continuous reactors, the ashes must be removed by diverting and collecting them near the wall as they diffuse outward from the center of the plasma.[q] Separating out the ashes is therefore more difficult in a continuous reactor since the separation must be done near the reaction region in the presence of high neutron flux and high temperatures. The separation of the ashes can be achieved, in principle, with a somewhat tricky magnetic field arrangement called a "divertor." The divertor scrapes off the outside layer of the plasma and guides it, by means of magnetic field lines, out of the main reaction region, where it is neutralized, collected, and purified for re-use as fuel.[r]

Conceptual Design of a D-T Reactor

Introduction

We consider now the design of a steady state, toroidal geometry, fusion reactor fueled with a 50/50 mixture of deuterium and tritium. For the design, little knowledge about details of the plasma is necessary to determine the approximate size and general features of the reactor. In fact, the design considerations for a D-T reactor turn out to be dominated by the behavior of the reactor materials that are subjected to the large flux of 14-MeV neutrons. The design of a reactor that operates on a reaction (such as D-He^3) whose energy appears as kinetic energy of charged rather than neutral particles would therefore be considerably different.

The design considered here follows, in large part, that of Carruthers, Davenport, and Mitchell [7], although we have chosen wall materials and power levels more in line with recent thinking. Their conceptual design is shown in Figures 2–3 and 2–4. The blanket configuration shown in Figure 2–3 is typical of the blanket designs for steady state reactors whether they are linear (using magnetic mirrors) or toroidal [8].

The magnetic field for confining the plasma is generated by superconducting coils placed on the outside of the blanket as far as possible away from the heat and the neutrons generated by the plasma. Placing the coils on the outside means that the magnetic field must be maintained in a very large volume. High magnetic fields in large volumes mean high cost, but the alternative of trying to maintain a superconducting magnetic coil ($T <$ 20°K) near the plasma ($T > 10^8$ °K + neutron flux) seems impossible.

A shield to protect the superconducting magnets from neutrons and x-rays is placed just inside the coil. Just inside the shield is the primary attenuator for moderating and absorbing the neutrons (converting their kinetic energy to heat), breeding tritium for fuel, and transferring the heat to some working fluid for use in a thermal cycle. In this particular design a fused salt, Li_2BeF_4 (sometimes called FLIBE), is used both as the primary attenuator and as the heat transfer fluid. Other, more recent, designs have used liquid lithium metal instead [9]. In either case, the heat is removed from the reactor by circulating the liquid coolant (neutron absorber) through the blanket and then to a heat exchanger. The tritium breeding, as mentioned before, is accomplished by slow neutron reactions with the lithium.

The niobium wall (first wall) is used to separate the liquid absorber from the vacuum chamber for the plasma. Niobium is chosen because it has good high temperature properties and is relatively transparent to 14-MeV neutrons and should hopefully undergo relatively little structural damage. In fact, as we mentioned before, no one knows just what will happen to niobium (or anything else) when it is subjected to a large flux of 14-MeV neutrons. Other materials suggested for use as the first wall include molybdenum, vanadium, stainless steel, and graphite. In addition to considering resistance to structural damage when choosing a material for the first wall, it is also important to choose a material whose atoms (1) are not easily dislodged from the wall and taken in as impurities by the plasma, and/or (2) have low Z so that radiation is minimized if they do get into the plasma.

A sobering thought is that all of the energy extracted from the reactor for use in a thermal cycle must pass through this wall as 14-MeV neutrons.[s] Moreover, the first wall is subjected to bremsstrahlung and synch-

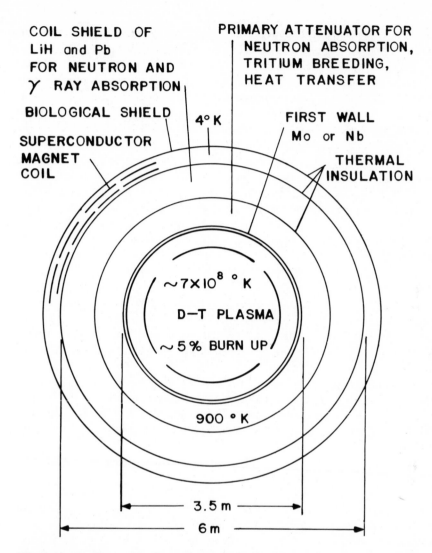

Figure 2–3. Cross Section of a D-T Fusion Reactor (after Ref. [8]).

rotron radiation from the plasma as well as to gamma rays produced by neutron irradiation of the structural materials (Mo, Nb, V, etc.) in the blanket. It is not hard to see that the first wall of a D-T reactor presents what is probably the most critical materials problem in the entire reactor design. Again we stress that if reactions were used that did not give off as much energy in the form of high energy neutrons, then the first wall materials problems would clearly be reduced. Unfortunately, it will be very

38

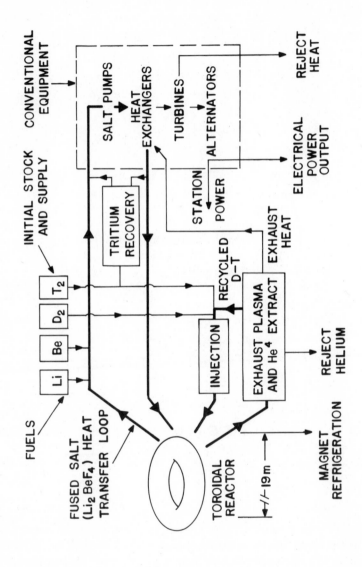

Figure 2–4. Conceptual Fusion Reactor System (after Ref. [7]).

difficult to achieve plasma temperatures high enough for these other reactions and to confine these plasmas long enough to break even on the energy.

We now proceed to estimate the physical size and electric power output of a D-T reactor. As we mentioned earlier, these quantities for a D-T reactor are determined primarily by properties of the blanket materials rather than by plasma properties.

Wall Loading

We define wall loading P_W as the reactor thermal output power divided by the area of the wall facing the plasma (the first wall) [7].[t] The maximum wall loading is limited by heating of the wall. This heating is caused by absorption of 14-MeV neutrons passing through the wall, of neutrons from the reactions which breed T in the blanket, of bremsstrahlung and synchrotron radiation from the plasma, and of gamma-ray backshine caused by activation of the blanket structural materials by 14-MeV neutrons. It turns out that the dominant heating mechanism is gamma-ray absorption. This heating is reduced as the wall is made thinner. On the other hand, the wall becomes structurally weak as it is made thinner. We assume a wall thickness of 0.25 cm and a wall loading for niobium of about 4 MW/m², typical "current" values.[u]

The rate at which energy is deposited in the wall at this wall loading is about 1/8 MW/m² due to the neutron flux and about 1 MW/m² due to radiation and charged particle bombardment [10].

Wall Radius

Using the notation of Figure 2–5, we see that the power per unit volume of the reactor P_D is given by the power produced in a short length ℓ of the torus, $2\pi r_W \ell P_W$, divided by the volume of the short length of torus, $\pi(r_W + t + s)^2 \ell$ [7]:

$$P_D = \frac{(2\pi r_W)\ell}{\pi(r_W + t + s)^2\ell} P_W \tag{2.6}$$

The value of $t + s$ can be considered to be independent of r_W, the first wall radius, since t, the thickness of the blanket, is primarily determined by how thick it must be to absorb all the neutrons, while s, the thickness of the magnet coil layer, is mainly determined by how much thermal insulation is required to insulate the superconducting windings from the hot blanket.

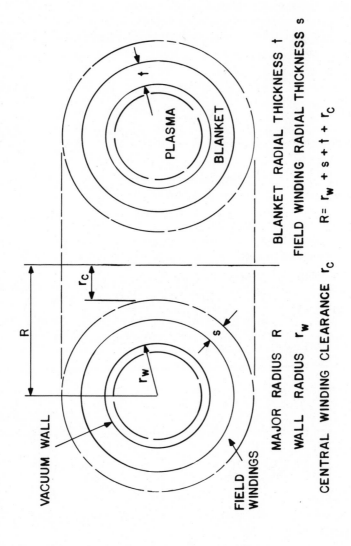

Figure 2–5. Toroidal Reactor Cross Section (after Ref. [7]).

If $t + s$ is a fixed number (determined by properties of materials), then the power density P_D can be maximized with respect to r_W. Maximization of P_D (for a fixed wall loading P_W) reduces the volume of magnetic field required for a given power output and hence reduces the capital cost of the magnet system, a very expensive item. By setting $dP_D/dr_W = 0$, we find that the maximum of P_D occurs for $r_W = t + s$. Following Caruthers et al., we take $t = 1.25$ m and $s = 0.50$ m. Thus, the best value for r_W is $t + s$ = 1.75 m, independent of P_W.

Power Density

Using the values of r_W, s, t, and P_W determined above, the average power density in our fusion reactor as given in equation (2.6) is

$$P_D = \frac{2\pi r_W}{\pi(r_W + t + s)^2} P_W = 1.1 \text{ MW/m}^3$$

This compares with a power density of 2–2.5 MW/m³ for advanced designs of fast fission breeder reactors. Thus, our power density does not seem to be unreasonable. The average power density calculated by using only the volume inside the reactor first wall rather than the entire volume of the torus is 4.5 MW/m³. For a practical value of the plasma radius, the power density in the plasma (thermal power out/volume of the plasma) is about 9 MW/m³. This is not greatly different from the 40 MW/m³ we talked about earlier and which we stated was actually somewhat too high for a continuous reactor.

Reactor Output

The area of the first wall in our toroidal reactor is given approximately by $(2\pi r_W)(2\pi R)$. Thus, the thermal power output for a wall loading of P_W is [7]:

$$P_T = 4\pi^2 r_W R P_W \tag{2.7}$$

Referring to Figure 2–5, we see that P_T has a minimum when $r_C = 0$ so that R is smallest. However, it is practically impossible to design a satisfactory magnet system with $r_C = 0$.

We suppose a minimum practical value of r_C to be 4.0 m in order to allow room for various fuel injection ports, vacuum ports, coolant pumps, and so forth. Using this value, we see that $P_T \approx 2100$ MW of thermal power. Assuming a typical thermal conversion efficiency of approximately forty percent, the electric power output is seen to be 840 MW(e), not unreasonably large by present or near future standards (see Appendix 2–B).

Table 2–2
Summary of Anticipated Reactor Parameters

	Parameters from Previous Discussion	Range of Parameter Values from Ref. [11]	Median Parameter Values from Ref. [11]
r_W	1.75 m	1.0–2.0 m	1.75 m
$r_W + s$	3.5 m	2.1–3.9 m	3.5 m
R	7.5 m	3.5–12.7 m	5.0 m
$P(th)$	2100 MW	2750–6000 MW	5350 MW
$P(e)$	820 MW	1170–2300 MW	2150 MW
n	10^{20}–10^{21} m^{-3}	3–4.9×10^{20} m^{-3}	4×10^{20} m^{-3}
T	10 keV	10–20 keV	20 keV
B	7 T	4–17 T	6.5 T
$n\tau$	$> 10^{20}$ m^{-3}	$(1.7$–$14) \times 10^{20}$ m^{-3}sec	4.5×10^{20} m^{-3} sec
$\eta = P(e)/P(th)$	40%	29.2–46.8%	40%
$\beta = 2nkT/(B^2/2\mu_0)$	0.06–0.11	0.043–0.4	0.12
Power density	4.6 MW/m^3	18.6–87 MW/m^3	29.4 MW/m^3

Summary of Reactor and Plasma Parameters

In the preceding paragraphs we have used rather simple arguments to estimate various plasma and reactor parameters in a D-T reactor. These parameters are summarized in Table 2–2. For comparison we list the corresponding parameters for Tokamak and toroidal reactors from several, more detailed, early fusion reactor studies that have been summarized by Persiani et al. [11].

The parameter values obtained here are clearly good "ball park" estimates of some important plasma and reactor parameters. Our reactor output power and power density are below the median for the other studies, however, because we have assumed lower wall loading P_W than in the earlier studies. This assumption is in line with more recent studies [11].

Laser Fusion and Inertial Confinement of a Plasma

Conceptually, perhaps the simplest way to produce thermonuclear fusion is to begin with a low temperature, solid fuel pellet and heat it to fusion temperatures so quickly that a large number of nuclei collide, fuse, and release energy before the plasma generated from the pellet has time to expand appreciably. With this approach, the nuclei are "confined" by their own inertia while they fuse. As a consequence, expensive confining magnetic fields are not required. As we can imagine, this inertial confinement time τ is small. Consequently, dumping enough energy into the pellet to heat it to fusion temperatures during the confinement time τ is very difficult. In the hydrogen bomb, which uses inertial confinement, the heating is accomplished by compression from a fission bomb detonation. In proposed inertial confinement schemes for controlled thermonuclear fusion, a pulsed, high power laser [12] or an electron beam [13] is intended to perform this task.

Another difficulty that results from the small value of τ for inertial confinement is in satisfying the condition[v]

$$n\tau > (n\tau)_{\text{break even}}$$

even when n is the value for a solid. As a result, the idea of compressing the pellet to super high density, perhaps several thousand times normal solid density, becomes attractive [14]. In laser fusion, this compression would be achieved by a kind of rocket action by particle "blow off" at the surface of the pellet as several (perhaps as many as 16–20) laser beams hit it from all sides (spherically symmetric irradiation). The required compression can be realized with lasers of reasonable size only if the shape of the laser pulse is carefully tailored and radial variations in the composition of the pellet are introduced. Under these favorable conditions, it seems that a laser system (multiple beam) that can deliver at least 10^4 J of energy in about, or less than, $10^{-3}\mu$sec will be required to demonstrate "scientific feasibility."[w] Presently, CO_2 laser systems that can deliver about 10^3 J in 10^{-3} μsec have been built at Los Alamos Scientific Laboratory and glass laser systems that can deliver 10 kJ in less than 10^{-3} μsec are being built (SHIVA laser facility at Lawrence Livermore Laboratory). In passing, we note that the shortness of the fusion pulse helps prevent instabilities from affecting the ion confinement. Instabilities among the lighter electrons, however, play an important role in transferring the laser energy to the ions to heat them.

Perhaps the major unanswered questions about laser fusion concern whether or not self-shielding by the plasma will prevent sufficiently fast energy transfer to heat the plasma; whether or not plasma instabilities will upset the compression and confinement of the pellet plasma; the cost, efficiency, and reliability of the lasers required; and the problem of designing reactors to withstand the "microbomb" explosions over an extended period of time.[x]

Fusion with Relativistic Electron Beams

Proposals have also been made to use relativistic electron beams instead of lasers to heat solid pellets to fusion temperatures [13, 15, 16]. This approach actually uses a combination of compression and inertial confinement (in the early stages of the fusion pulse) and magnetic confinement (later on). The magnetic confinement arises from the extremely strong (several hundred Tesla) magnetic field produced by the relativistic beam itself. The advantages of relativistic electron beams, in comparison with lasers, include their high efficiency (> 40%) and the fact that electron guns with an output energy greater than that theoretically required to reach the break-even point have already been manufactured (although not with quite the right parameter combinations). The basic difficulty at present is to focus the beam to a sufficiently small spot to get the high energy *density* required to reach the break-even threshold.

Costs

Capital costs and fuel costs for power stations are obviously very important considerations since they represent an appreciable part of the cost of supplying electric power. Compared to conventional fossil fuel power plants, fusion power plants are expected to have negligible fuel costs but high capital costs. In a fusion reactor, therefore, the cost of the electrical power will be largely determined by the capital cost of the reactor. Unfortunately, the capital cost of a fusion reactor is difficult to determine at this time. Many of the design details are unknown. Those that are known involve developing technologies (such as superconducting magnets) whose costs are dominant factors in the capital costs but for which the ultimate production costs have not yet been determined.

Perhaps more important than determining the absolute capital cost is determining the cost of generating power relative to fast fission breeder reactors, which are likely to be a common kind of power plant when nuclear fusion plants are ready to come on line—perhaps near the turn of the century. About all that can be said at present is that there seems to be no reason why nuclear fusion reactors could not produce power at costs comparable to those of fast fission breeder reactors.

If the additional costs of making nuclear reactors (fission or fusion) acceptable from an environmental and safety point of view are considered, then nuclear fusion reactors seem even more likely to become economically competitive with fast fission breeder reactors. For one thing, there is no theoretical possibility of a nuclear excursion (jargon for nuclear runaway conditions in the reactor) in a nuclear fusion reactor (there is not enough fuel in it at one time to permit an H-bomb type explosion) as there is in a fast fission breeder. This feature of nuclear fusion reactors, together with the absence of air pollution, might make it possible to locate them directly in population centers where the power consumption is highest and thus permit savings in distribution costs. It seems that even in D-T reactors, in which there are definitely radioactive materials, the radioactivity is much less severe than in fast fission breeders.[y] Thus, it seems reasonable that the costs of safety precautions should be less for fusion reactors than for fission reactors. Perhaps more important in the (very) long run is the possibility of building a nuclear fusion power reactor completely free of radioactivity by using a reaction such as H-Li^6 for which no neutrons are produced. (In D-He^3 reactors, D-D reactions would still produce a few neutrons which would induce some radioactivity in the reactor materials.) If such a reactor became available, it would be almost the ultimate power source—being practically free of pollution of any kind (radioactivity, heat, etc.).

Notes

(a) More careful fusion reactor design considerations give a value below 40 MW/m^3 for continuous reactors and above it for pulsed reactors (during the peak of the pulse).

(b) This temperature is sometimes called, as in Ref. [6], the ideal ignition temperature. We reserve the term ignition to describe a somewhat different operating regime. For a discussion of this point as well as how the ideal break-even temperature might be reduced see Appendix 2–A.

(c) Notice that the Lawson Criterion does not correspond to ignited operation (Appendix 2–A) of the reactor since it assumes part of the output power is recirculated to heat the plasma.

(d) For ignited operation, a larger $n\tau$ is required.

(e) Calculations for the two ion component scheme mentioned in an earlier note indicate that break even with D-T might be reached with $n\tau \approx 10^{19}$ m^{-3} sec. This reduction in the required $n\tau$, which corresponds to an order of magnitude reduction in confinement time, is the main attraction of the two ion component approach. The decrease in $n\tau$ is realized, as mentioned before, by obtaining fast ions by injecting them into the plasma rather than by heating the entire plasma to high temperature, T. Thus the energy required to "heat" the plasma is less than the $3nkT$ used in deriving the Lawson criterion. This advantage is achieved at the expense of increased circulating power. This added circulating power will probably prevent the use of the two ion component scheme in a commercial reactor for economic reasons. (Note the discussion in Appendix 2–A concerning lowering the ignition temperature by increasing the circulating power.) The two ion component scheme can be useful for building a reactor to produce 14-MeV neutrons for testing materials, however (see discussion in a later section).

(f) The ratio of kinetic pressure in the plasma, $2nkT$, to the magnetic pressure outside the plasma, $B_0^2/2\mu_0$, is usually called the "beta" of the plasma: $\beta = 4\mu_0 nkT/B_0^2$. If $\beta \lesssim 0.1$, the plasma is said to be "low beta." If $\beta \gtrsim 0.1$, it is said to be "high beta."

(g) Although the part of the magnetic field that penetrates the plasma is lost as far as confinement is concerned, it does permit a certain amount of control of the magnetic field configuration (shear, minimum-B) within the plasma, from outside. Thus, the magnetic field can be shaped to help prevent instabilities, for example. In some fast pulsed experiments, the magnetic field does not have enough time to penetrate the plasma during the plasma lifetime ($\beta \approx 1$). Such experiments make good use of the magnetic field for confinement, but, of course, permit less direct control of the magnetic field within the plasma, which can be a disadvantage from the stability standpoint, since shear and minimum-B concepts are not directly applicable. Dynamic and feedback stabilization can, therefore, be particularly important in such cases.

(h) The Alcator experiment at MIT.

(i) One other possibility is to use the high energy neutrons released by fusion to transmute $_{92}U^{238}$, which is not fissionable into $_{94}Pu^{239}$, which is fissionable and could be used as fuel in a fission reactor. It turns out, in fact, that a fusion reactor might be so effective in generating fuel for a plutonium-fueled fission reactor that it is not even necessary to reach the Lawson criterion in the fusion reactor in order to realize a net power output from such a symbiotic combination of fission and fusion reactors. Fission-fusion symbiosis or a "hybrid reactor" (see L. M. Lidsky, "Fission-Fusion Systems: Hybrid, Symbiotic and Augean," *Nuclear Fusion* 15, pp. 151–173, February 1975) has the potential advantages of (1) providing an alternative means to breeder reactors for utilizing the energy available from vast reserves of U–238 and (2) providing design and operating experience with fusion reactors without satisfying the difficult energy break-even constraint. Its disadvantages include increased complexity (and hence cost) and a combination of the disadvantages of both fission and fusion. The lack of past interest in fission-fusion symbiosis probably results more from the tendency of a nuclear community traditionally divided into fission and fusion camps to view symbiosis as a half-breed rather than from its disadvantages. Recently there appears to have been a renewed interest in the hybrid reactor scheme, however.

It has also been proposed to use the high energy fusion neutrons to transmute the highly radioactive elements in the waste from fission reactors into less radioactive elements which could be more easily stored (see, for instance, G. Kaplan, *IEEE Spectrum* 12, No. 9, p. 62, September 1975).

(j) For a discussion of direct energy conversion see George H. Miley, *Fusion Energy Conversion* (American Nuclear Society, Hinsdale, Illinois, 1976), chapters 3 and 4.

(k) At the time of writing, ERDA plans to have a source with a maximum flux of 1×10^{13} neutrons/cm^2-sec and a test volume of 1 cm^3 operating in 1978. A source with a maximum flux of 1.2×10^{14} neutrons/cm^2-sec and a test volume of 3 cm^2 should be operational in 1981.

(l) Notice that one triton (tritium nucleus) produces one neutron, which in turn can produce at most one triton by being captured by a Li^6 nucleus. Actually, some neutrons will be lost by capture in other kinds of nuclei. In order to prevent the tritium inventory in the reactor from gradually decreasing, neutron multiplication reactions must be used in the blanket. One possible reaction is $_4Be^9 + _0n^1 \rightarrow 2\,_2He^4 + 2\,_0n^1 - 1.67$ MeV.

(m) Fortunately, only relatively small amounts of tritium are required. Using the fact that each D-T fusion releases 17.6 MeV of energy and each n-Li^6 reaction releases 4.6 MeV, it is easy to show that a 2000-MW (electrical) plant operating at 40% thermal efficiency would require, very roughly, a kilogram of tritium per day. This rate of use would probably require a tritium inventory of only several kilograms to provide sufficient time for tritium to be recovered from the blanket and made available for use as fuel.

(n) The two ion component scheme mentioned in earlier notes is an example of unignited operation.

(o) In mirror machines, the high operating temperature (discussed in a later section) in conjunction with the large refuelling rate required to compensate for end losses, may require high energy neutral beams intense enough to maintain (refuel) the reactor plasma, as well as to heat it.

(p) Heating by compression occurs in the following way. As the confining magnetic field is increased, the magnetic pressure on the plasma also increases. This increased pressure causes the particles to move toward the interior of the plasma, and the plasma contracts. The kinetic energy of this inward motion of the particles is soon randomized by collisions with plasma particles. The energy then appears as increased thermal energy of the plasma.

(q) Some theoretical calculations indicate that the impurities may diffuse in toward the center of the plasma. If so, the ash removal will be more difficult and radiation losses could be substantially increased.

(r) For a discussion of divertors, see George H. Miley, *Fusion Energy Conversion* (American Nuclear Society, Hinsdale, Illinois 1976) pp. 139–144.

(s) This is true of course, only if the energy carried by the alpha particles is used to heat the new fuel, or for some other reason is not converted directly to increase the electric power output of the reactor.

(t) This definition assumes that none of the alpha-particle energy is converted directly to output power from the reactor.

(u) The assumed wall loading in fusion reactor design has been decreased by nearly an order of magnitude over the past few years as a better understanding has been reached of the severe materials problems.

(v) See Appendix 2–C.

(w) $n\tau \geq 10^{20}$ m^{-3} sec and $T > T_{ideal}$. See Appendix 2–C.

(x) The major motivation for present laser fusion work, however, may not be to produce fusion power but to use the microbomb explosion as a substitute for the fission bomb trigger in the hydrogen bomb or to simulate radiation effects from nuclear weapons.

(y) Tritium, probably the most volatile radioactive element in a fusion reactor, has a "biological" half life of only several days and hence passes through organisms fairly quickly and without tending to collect in bones, etc., as some fission wastes do. This circumstance, together with the fact that tritium emits only beta particles (electrons) at a relatively low energy, works to reduce substantially the potential biological radiation damage from the tritium in a fusion reactor in comparison to that from the fuels and wastes in a fission reactor. Other radioactive wastes, such as discarded reactor components (radioactive because of neutron bombardment), represent only a small fraction of the fission fragment wastes from a nuclear fission plant.

Appendix 2–A

Break Even and Ignition in Fusion Reactors

A reactor can actually deliver useful power with temperatures less than the ideal break-even temperature if the escaping radiation is collected, converted into electricity, say, and some part of it is used to heat the plasma. This procedure clearly reduces the net cooling effect of the radiation on the plasma and hence permits radiation losses to be overcome at a temperature lower than the ideal break-even temperature. If the break-even temperature is to be lowered appreciably, however, a considerable part of the power radiated as bremsstrahlung must be recirculated as described above. Recirculating this power can substantially increase the capital cost of a reactor. As a consequence, the total recirculating power probably should not exceed ten percent, say, of the reactor power output. At this point, it is important to note that the power released as neutron kinetic energy must also be recirculated to break even.

The neutrons (being neutral and hence not confined by the magnetic field) escape the plasma region and carry with them most of the fusion energy. Thus, they must be collected and their energy converted to electricity and used to drive a supplementary heating device that reinjects their energy into the plasma. (The alpha particles, being charged, can be confined by the magnetic field within the plasma where they give up their energy to the plasma particles by means of collisions.) It is thus unlikely that the break-even temperature in a D-T reactor can be reduced very much below its ideal value and yet have the total (radiation plus neutrons) recirculating power kept low. Recirculating power considerations similarly limit the applicability of the two ion component approach to fusion, as mentioned in an earlier note. Energy must be circulated continually to sustain the energetic beam of ions. Since recirculation of energy is costly, this approach will probably be used only in experimental or test reactors in which cost is not a dominant consideration.

Now we can describe what is meant by ignition in a reactor. Specifically, ignition occurs when the fusion reaction rate is large enough that the energy that appears in the alpha particles alone (without that of the neutrons) is large enough to offset the radiation loss from the plasma. The idea here is that the energy deposited in the plasma by the alpha particles (which can be confined by the magnetic field) is enough to keep the reaction going without the need to recirculate a part of the energy carried by the neutrons. Operating a reactor in the ignited mode could therefore be very desirable in that the capital cost of the reactor is reduced by the fact that no circulating power is necessary.

A potential disadvantage of ignited operation is that the reaction rate and temperature in the ignited mode are unstable to small perturbations away from their equilibrium values. This instability results from the fact that the rate of change of fusion power with respect to temperature is larger than the rate of change of radiation loss with respect to temperature. As a result, a small increase in plasma temperature produces an increase in both the fusion power released and the radiation loss rate but a net increase in their difference, the rate at which energy is deposited in the plasma. As a result, the original temperature perturbation is multiplied. Thus, the perturbation grows and becomes unstable. Nevertheless, the reactor cannot blow up like a bomb since there is only a relatively small amount of fuel in the reaction region at any one time. (In contrast, fuel for a few years operation is always present in the core of a fission breeder reactor.) It is conceivable, however, that the temperature could increase enough to damage parts of the reactor. It may be possible to stabilize an ignited reactor by, for example, using a control system to inject small amounts of high Z impurities to damp out an increase in temperature. In passing, we note that control of the reaction rate in an *unignited* reactor can be achieved simply by adjusting the level of supplementary heating supplied to the plasma.

Appendix 2–B

Size Considerations for Fusion Reactors

In order to support the claim that the properties of blanket materials dominate the design of a D-T reactor, we note that the wall loading P_W determines not only the reactor power output, but the plasma density n, plasma temperature T, and, indirectly, the confining magnetic field as well. To see how this comes about, consider a short sector of length ℓ of a toroidal reactor. The power produced in the sector is given by the thermonuclear power density of the plasma P_{th} times the volume of the plasma in the sector $(\pi r_p^2)\ell$:

$$P_{\text{sector}} = (\pi r_p^2)\ell P_{th}$$

where P_{th} is some function $g(n, T)$, as shown in Figure 2–1. The wall loading is P_{sector} divided by the wall area of the sector, $(2\pi r_W)\ell$. Thus

$$P_W = \frac{g(n, T)}{2} \frac{r_p^2}{r_W}$$

Now to make good use of the magnetic field volume, we must take $r_p \lesssim r_W$ so that

$$P_W \approx (1/2)g(n, T)r_W$$

But r_W has already been chosen to be $t + s$ to optimize the power density in the reactor. Thus, n and T must be chosen to achieve the proper value of P_W. If we make the reasonable assumption that there is some maximum plasma kinetic pressure which can be confined by the reactor (because of mechanical stress or technological or economic limits to the size of the magnetic field, for example), then we must have

$$p_{\text{max}} \geq p = k(n_e T_e + n_i T_i) \approx 2nkT$$

Thus, with such a constraint, the allowable particle density is $n \propto 1/T$. We note from Figure 2–1 that the thermonuclear power density $P_{th} = n^2 f(T)$, where $f(T)$ is some function of temperature. Subject to the above constraint, therefore,

$$P_{th} = \left(\frac{p_{\text{max}}}{2k}\right)^2 \frac{f(T)}{T^2}$$

Thus, for a given p_{max} (i.e., a given stress), we can get different thermonu-clear power densities at different temperatures as the factor $f(T)/T^2$ var-ies. It happens that for the D-T reaction, the function $f(T)/T^2$ has a maxi-mum at $T \approx 10$ keV (see, for example, R. G. Mills, *Nuclear Fusion* 1, 223, 1967). Thus with $T \approx 10$ keV, we maximize the thermonuclear power pro-duced for a given stress on the reactor. With the choice $T = 10$ keV, n must be chosen so as satisfy the limits on P_W and/or p_{max}. Thus, p_{max} and P_W effectively determine T and n. It is a happy circumstance that this tem-perature is in the range that gives the minimum $n\tau$ requirement (see the discussion of the Lawson criterion).

A second requirement on the plasma and the magnetic field is that the confinement time τ be large enough so that we can extract sufficient ener-gy from the plasma to keep the reactor going and have some left over to use. The required τ is consequently somewhat larger than that necessary to satisfy the Lawson criterion. The confinement time τ increases with magnetic field strength as we discussed previously, and with the size of the plasma (since a typical particle must travel further to escape). It also depends on the plasma temperature and density. If the plasma radius, density, and temperature have been chosen to give the proper P_W at the wall, however, then clearly the magnetic field must be chosen to give an adequate confinement time (output power) and hence is indirectly deter-mined by P_W. Since there are technological and economic limits to the magnetic field strength, it is most important to maximize τ by choosing the shape, and other parameters, of the magnetic field and the plasma to suppress instabilities and hence the consequent enhanced plasma losses. If all of this is done, we note, as previously mentioned, that the reactor design is dominated by blanket material considerations.

It is perhaps worth noting that a lower magnetic field strength could be used if the plasma radius were increased to a value larger than $t + s$ while the other plasma parameters were held fixed. The increase in plasma radi-us would necessitate a somewhat greater increase in wall radius to keep the power density at the wall equal to P_W, since the radiation output from a short length ℓ of the plasma is proportional to r_{plasma}^2 ($\propto$ plasma volume for fixed length ℓ) while the power density at the wall is proportional to $1/r_W$ ($r_W \propto$ wall area for fixed length ℓ). This larger value of r_W will de-crease the power density of the reactor and hence tend to increase its cap-ital cost; however, the resulting decrease in magnetic field strength tends to offset this increase in cost. Whether or not such a trade-off could be made profitably depends on the cost trade-offs between magnetic field strength and magnetic field volume.

Appendix 2–C

Break Even for Laser Fusion

A glance back at the relevant calculations shows that the $n\tau$ and T values are not appreciably different for magnetic and inertial confinement as long as (1) synchrotron radiation (which occurs only in the presence of a magnetic field) is negligible and (2) the overall energy conversion efficiency, taken as 1/3 earlier, is the same. The overall energy conversion efficiency is the efficiency with which energy released by fusion can be transformed back into thermal energy of the plasma to keep it hot. In D-T schemes, therefore, the overall energy conversion efficiency includes as a factor not only the efficiency with which the fusion energy (mainly neutron energy) can be converted to electricity, but also the efficiency with which electric energy can be used to heat the plasma. In magnetic confinement reactors, the efficiency of the plasma heating devices may be relatively high—perhaps as high as eighty percent. Thus an overall conversion efficiency of 1/3 is not altogether unreasonable. In laser fusion devices, however, the efficiency with which electricity can be converted into plasma heat is no greater than the laser efficiency, which ranges from less than 0.1% for Nd-doped glass lasers to a theoretical maximum efficiency of about 40% (actually much less ($< 10\%$?) for pulsed CO_2 lasers). Thus the overall energy conversion efficiency would be about 0.04% for Nd-doped glass lasers and maybe 5% for CO_2 lasers if we assume that the fusion energy can be converted into electricity with an efficiency of 40%. Even these low figures assume complete absorption of laser energy by the plasma. The lower overall energy conversion efficiencies therefore increase the $(n\tau)_{\text{break even}}$ for laser fusion above the value of 10^{20} m^{-3}-sec usually quoted for magnetic confinement. From these considerations, it is clear that CO_2 lasers have considerable advantage over Nd-doped glass lasers for fusion purposes as far as efficiency is concerned. The major problems with CO_2 lasers are packing enough energy into a short enough pulse and absorbing the pulse in the plasma. It appears more difficult to obtain a high absorption coefficient for long wavelength laser, such as CO_2, in such a scheme. Even the most ardent supporters of laser-pellet fusion admit that the Nd-doped glass laser system is not the solution for a successful fusion reactor system. Most of them agree that a gas laser system is needed, since the medium is self-repairing (to optical damage) and since direct pumping methods, such as electron beams are possible (flash-lamp pulsed lasers are inherently low-efficiency). It is also generally felt that

53

the CO_2 laser, which is reasonably efficient, has too long a wavelength for efficient absorption. The "ideal" laser is thus one with an active volume of gas, with high efficiency and a short wavelength (in the visible wavelength range). This, as of yet undiscovered, laser is commonly denoted as the "Brand X laser."

As a final point, we note that for a fixed overall energy conversion efficiency, the minimum $n\tau$ and T values for inertial confinement are more favorable than those for magnetic confinement if synchrotron radiation is not negligible in the magnetic case.

**Part II
Fusion Devices**

Introduction to Part II

We now turn our attention from fusion in general to particular approaches to fusion. Since there are lots of different approaches, we focus primarily on the ones presently receiving significant support from the USERDA. We devote particular attention to the Tokamak since it is presently given top priority by ERDA and serves as a convenient vehicle for exploring many aspects of plasma physics relevant to the operation of other devices. More abbreviated discussions of mirror machines, theta-pinches, and laser pellet fusion are included. These four devices represent the main thrust of the US fusion program. There are those who feel that this program is shortsighted in its emphasis on plunging headlong toward the goal of a D-T low beta Tokamak reactor, which may offer relatively few of the advantages often attributed to controlled thermonuclear fusion [17, 18, 19]. Some feel that more emphasis should be placed on alternative concepts that have a better chance of ultimately burning one of the harder to ignite but more desirable fuel cycles.[a] These include those mentioned above plus several others which have not yet been investigated in as much detail. Because of the possible future importance of such devices, several of these are briefly described in a concluding section.

---

[a] Refs. [17], [18], and [19] constitute a useful critical review of the USERDA program in controlled fusion.

3

Tokamaks

Introduction

We have already seen that toroidal field configurations offer the advantage in plasma confinement of having no "ends" through which the plasma can escape. Furthermore, we saw that if the magnetic field lines are made to spiral around the torus, the drift of particles across the magnetic field averages to zero. In short, a magnetic field whose lines are toroidal helices seems to be a good candidate for confining a plasma. How can we, in practice, realize such a field configuration? As shown in Figure 1–4a, field lines with no spiraling can be produced by simply passing current through a coil wound on a toroidal form. The trick is to get these lines to spiral as they pass around the torus. One way (but not the only one) to do it is just to pass a current through a plasma (an excellent conductor) located inside the toroidal coil. The field produced by the plasma current alone is shown in Figure 3–1. This field, when superimposed on the field produced by the toroidal coil, gives a magnetic field whose lines are toroidal helices, the desired field configuration. In terms of a toroidal coordinate system the toroidal coil produces a magnetic field B_t in the toroidal (or ϕ) direction, while the plasma produces a field B_p in the poloidal (or θ) direction. A toroidal confinement device such as the one just described in which the toroidal component of the magnetic field is produced by poloidal currents in external coils while the poloidal field is produced by a toroidal current in the plasma is called a Tokamak [20, 21].[a] The basic Tokamak machine is shown in Figure 3–2. Actually, the toroidal and poloidal fields alone are not enough to confine the plasma in a Tokamak. The plasma current interacts with the poloidal field (produced by the plasma current), which is weaker on the outside of the torus than on the inside, to produce a net $\mathbf{J} \times \mathbf{B}$ force outward along the major radius of the torus. The result is that the plasma escapes simply by increasing its major radius. Equilibrium thus can be established in a Tokamak only by applying a restoring force to overcome this tendency of the plasma to expand outward and hit the wall. The restoring force can be applied in two ways, as shown in Figure 3–3. First, a "vertical" magnetic field B_v can be applied to the plasma in a direction parallel to the major axis of the torus (Figure 3–3(b)). This magnetic field interacts with the toroidal plasma current to produce inward directed $\mathbf{J} \times \mathbf{B}_v$ force which stops the outward expansion

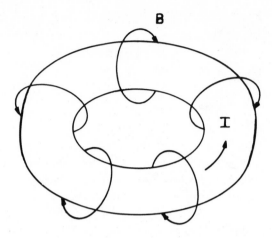

Figure 3–1. Field Lines Produced by Toroidal Current.

of the plasma. The toroidal plasma current is strong enough so that only a small B_v is required to achieve equilibrium. In fact, a strong enough B_v can be produced on a transient basis by eddy currents induced in a toroidal copper "stabilizing" shell placed around the plasma (see Figure 3–2). As the plasma tries to expand outward along its major radius R the poloidal field lines "cut" the copper shell and induce a toroidal current in its outer edge. This current then produces a primarily vertical field in the plasma region (see Figure 3–3(a)). The copper shell technique was used exclusively in early experiments, while in later experiments, vertical fields applied by external coils often have been used so that the radial position of the plasma in equilibrium can be controlled by adjusting the size of B_v.

The stainless steel liner shown in Figure 3–2 serves as a vacuum chamber of relatively low electrical conductivity that magnetic fields can easily penetrate. The laminated core transformer in Figure 3–2 has the plasma as its secondary so that the toroidal current required to produce the poloidal field can be induced in the plasma by applying a voltage to the primary windings. As a sort of bonus, the plasma is ohmically heated by the toroidal current. Unfortunately, the resistance of the plasma, and hence the heating rate, decreases as the plasma heats up.[b] The ohmic heating rate (which decreases with T) typically drops to the plasma heat loss rate (which increases with T due to radiation, conduction, etc.) at a temperature much less than that needed in a reactor. Thus ohmic heating alone will not suffice in a Tokamak reactor.

An important feature of Tokamaks is that their magnetic field configuration has shear built into it and is, in an average sense, a minimum-B configuration as well (see chapter 1, "Magnetic Confinement of a Plasma").

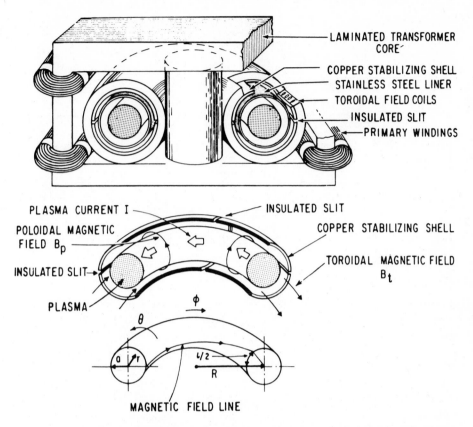

Figure 3–2. Basic Tokamak Apparatus: a toroidal plasma confined in a helical magnetic field created by the superposition of a strong, externally generated toroidal field and the poloidal field generated by the plasma current. The plasma current, induced by transformer action, resistively heats the plasma (after Ref. [20]).

Shear occurs because the plasma current produces a poloidal field which is small near the minor axis of the torus and increases to a maximum near the edge of the plasma[c] while the toroidal field varies slowly across the cross section. The net result is a change in the pitch of the helical field lines as we move away from the minor axis of the torus—that is, the magnetic field is sheared. The average minimum-B property also follows from the fact that the poloidal field is small near the minor axis and increases near the plasma edge so that the strength of the magnetic field, $B = (B_t^2 + B_p^2)^{1/2}$ has a minimum near the minor axis. Portions of each field line, however, pass through regions that are not in the magnetic well near the

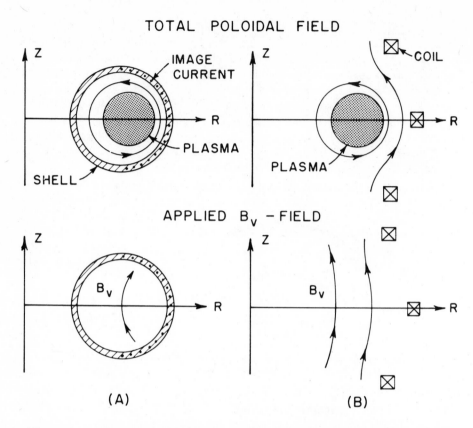

Figure 3–3. Plasma Equilibrium in Major Radius. This can be provided with a copper shell (A) or by external coils only (B). In either case, the required inward force (along *R*) is due to the Lorentz force from plasma current flowing across the "vertical field" B_v. In case (A), the B_v field is produced by image currents in the copper shell; in case (B), the B_v field is externally applied. Total poloidal field patterns are shown above; B_v field patterns by themselves are shown below (after Ref. [20]).

minor axis. As they move around the torus, the particles, which tend to follow field lines, therefore spend a part of their time inside the magnetic well and a part of their time outside it. It turns out that in the Tokamak field configuration, they spend more time inside the well, on the average, than on the outside. The Tokamak is therefore said to be an average minimum-*B* device. This property, together with shear, makes the Tokamak stable to most MHD or macroscopic instabilities in normal operating regimes.

Another important feature of Tokamaks is their relative simplicity of construction. More specifically, the relatively complicated magnetic field configuration required for toroidal confinement is produced by using a very simple coil system to produce the toroidal field and using a transformer to induce the toroidal plasma current which not only produces the relatively hard-to-come by poloidal component of the magnetic field, but also heats the plasma.[d] The operation of Tokamaks, on the other hand, is relatively complicated. In fact, the complexity results from the fact that changes in plasma parameters can affect the toroidal current, which can then change the poloidal field, which in turn can influence confinement and hence the plasma properties. Thus, the Tokamak equilibrium is, by nature, more complicated than the equilibrium for devices in which the confining fields are produced solely by external coils and are hence subject to direct external control that is more or less independent of the plasma.

The most important feature of Tokamaks, however, is not their simplicity of construction or their complexity of operation. Rather, it is that they seem, at the present time, to work better than other plasma confinement devices and to be scalable to reactor sizes. The theory of operation and the scaling laws have been reasonably well developed and are described in the following sections. Much of the theory and most of the conclusions have yet to be verified experimentally, however, and some existing data are not completely understood. Even so, Tokamaks are probably better understood than any other plasma confinement devices with the possible exception of mirror machines and linear theta pinches.

MHD Properties of Tokamaks

We now wish to examine the MHD stability of Tokamaks in a little more detail in order to see how the onset of MHD instabilities determines the permissible range of plasma density and current in Tokamaks and also determines the maximum β.

To begin our consideration of MHD stability of Tokamaks, recall our earlier discussion of the Kruskal-Shafranov limit in which we saw that if the pitch of the helical field lines were such that a spiral is completed during a single trip around the torus, then instability would occur. That is, a necessary, but not sufficient, condition for stability was to require the rotational transform ι (iota) to be less than 2π. Now ι clearly depends on B_p, which in turn depends on the plasma parameters through the plasma current. Thus, we need to find the relationship between ι and B_p in order to investigate the range of plasma parameters and currents that permit MHD stable operation of the Tokamak.

First, recall that the magnetic field **B** can be written as

$$\mathbf{B} = B_t \hat{a}_\phi + B_p \hat{a}_\theta \tag{3.1}$$

where $\hat{a}_\phi$ and $\hat{a}_\theta$ are unit vectors in the toroidal and poloidal direction, respectively, $B_t = B_t(r, \theta)$ is the toroidal field, $B_p = B_p(r, \theta)$ is the poloidal field, and r is the radial coordinate measured from the minor axis of the torus (see Figure 3–2).[e] The pitch angle of the magnetic field line with respect to the minor axis of the torus is γ, where $\tan \gamma = B_p/B_t$. During a single trip around the torus (the long way), the line moves a distance $2\pi R \tan \gamma = 2\pi R (B_p/B_t)$ in the poloidal direction. This distance is precisely the same as $\iota 2\pi R$. Thus,

$$\iota = 2\pi \frac{R}{r} \frac{B_p}{B_t}$$

Sometimes, it is convenient to talk about the variable,

$$q \equiv \frac{2\pi}{\iota} = \frac{r}{R} \frac{B_t}{B_p}$$

which is how many transits a line of force makes around the torus in the ϕ direction in making a single transit in the θ direction. Thus, the Kruskal-Shafranov limit can be written as

$$\iota < 2\pi \tag{3.2}$$

or

$$q > 1$$

Since it indicates by what factor the Kruskal-Shafranov criterion is satisfied, q is usually called the safety factor. In general, q varies over the plasma cross section. It thus is often convenient to express q at the outer edge of the plasma. This value of $q(r)$ is usually denoted $q(a)$. More detailed theoretical analysis indicate that a sufficient condition for MHD stability is something like[f] $q(a) \gtrsim 2.5$. This condition, which clearly limits B_p and hence the plasma current, is the first of two main results from the MHD stability analysis of Tokamaks we shall use. We will explore the implications of this limit more fully a little later on. First, however, we take a look at the second main MHD result: for an equilibrium to exist in a Tokamak, the poloidal magnetic field must be at least strong enough to support a fraction a/R of the kinetic pressure.[g] That is, one must require

$$\frac{B_p^2}{2\mu_0} \gtrsim \frac{a}{R} nk(T_e + T_i)$$

Notice that for $a/R = 0$, this result tells us something that we already know: in the limit of straight field lines ($a/R = 0$), no poloidal magnetic field is required to achieve an equilibrium. As the curvature of the lines is increased, however, the poloidal field must be increased so that the poloidal magnetic pressure can support more and more of the kinetic pressure. An interpretation of this result is that as the field lines are curved more and more, the gradient of the toroidal field, across the plasma, decreases the effectiveness of the toroidal field in confining the plasma. The confinement must therefore be shored up by the poloidal field as a/R increases.

It is customary to define the poloidal beta β_p as

$$\beta_p \equiv \frac{nk(T_e + T_i)}{B_p^2/2\mu_0} \tag{3.3}$$

Our second main MHD result can then be rewritten as[h]

$$\beta_p \lesssim R/a$$

where R/a is usually called the aspect ratio (of the Tokamak).

One of the main consequences of these two MHD results is a limit on the β in Tokamaks. Since β is the ratio of plasma kinetic pressure to the total magnetic pressure, β is a measure of how effectively the magnetic field in a plasma device is used to confine the plasma. If in a hypothetical plasma device we have $\beta = 0.01$, then the magnetic pressure is one hundred times the kinetic pressure. Such a device appears to make poor use of the magnetic field, since it seems that we should be able to confine a plasma with a magnetic pressure just equal to the plasma kinetic pressure ($\beta = 1$). Since large magnetic fields are hard to get, and even those that we can get are expensive, we clearly want β to be as large as possible in a reactor. The limit on β in a device is therefore a very important consideration.

To calculate the limit on the total β in a Tokamak, consider the ratio

$$\frac{\beta}{\beta_p} = \left(\frac{B_p}{B}\right)^2 = \frac{B_p^2}{B_t^2 + B_p^2} = \frac{1}{1 + (B_t/B_p)^2}$$

But

$$q \equiv \frac{a}{R}\frac{B_t}{B_p} \gtrsim 2.5$$

means that

$$\frac{B_t}{B_p} \gtrsim 2.5\frac{R}{a}$$

and $\beta_p \lesssim R/a$ so that

$$\beta \lesssim \frac{R/a}{1 + 6.25\left(\dfrac{R}{a}\right)^2} \lesssim \frac{a}{6R} \tag{3.4}$$

In our earlier reactor design calculation, we had $R = r_W + s + t + r_C$ where $r_W = s + t = 1.75$ m, $r_C = 4.0$ m, and $a = r_p \lesssim r_W = 1.75$ m. For that design, therefore, $a/R \lesssim 0.23$ and $\beta \lesssim 0.04$. If we assume somewhat more optimistically that $a/R \approx 1/3$, then $\beta \approx 0.05$. We are therefore forced to conclude that β_{max} for Tokamaks of the usual kind is limited to low values by fundamental MHD considerations.[i]

At this point we might naturally ask why in the world everyone is so fired up about Tokamaks when they are limited to such small values of β. The basic answer is that, as far as confinement is concerned, they work better than most other devices (and appear to be scalable to reactors of convenient sizes in a more or less straightforward manner). Ironically, one of the reasons that they work so well is their low beta. Recalling our earlier discussion of magnetic confinement, we know that in a high β plasma ($\beta = 1$) most of the applied magnetic field is excluded from the plasma so that it cannot be used to optimize confinement by providing shear and the like to prevent instabilities. The Tokamak β is low enough, however, to allow excellent control of the magnetic fields within the plasma and hence the Tokamak works well as a confinement device—but at the cost of low β.

In order to understand this trade-off between β and plasma stability a little better, recall that the minimum magnetic field ($\beta = 1$) required to confine a plasma with a kinetic pressure of $p = nk(T_e + T_i)$ is

$$B_{min} = \sqrt{2\mu_0 nk(T_e + T_i)} \tag{3.5}$$

In a device with a confining magnetic field B_c we can write the β as

$$\beta = \left(\frac{B_{min}}{B_c}\right)^2$$

so that the confining field required for a given β is

$$B_c = \frac{1}{\sqrt{\beta}}B_{min} \tag{3.6}$$

This relation is plotted in Figure 3–4. Notice that the magnitude of the magnetic field required to confine the plasma B_c drops sharply as β increases from 0 to about 0.1, but decreases much more slowly as β increases from 0.1 to 1. For example, $\beta = 0.1$ requires a confining field of only about 3.2 times that for $\beta = 1$ (optimum utilization of the magnetic field) and, of course, $\beta = 0.1$ permits considerable external control of the magnetic field within the plasma. If from our earlier design calculation for a D-T reactor we use $B_{min} = 2.0$ T, then for $\beta = 0.1$, we must have $B_c = 6.4$ T—a value that is not technologically unreasonable. The point is, therefore, that when the potential stabilizing effects of low β are weighed against the reduction in B_c for high $\beta(\beta \to 1)$, then $\beta \approx 0.1$ may be a reasonable operating condition. Even $\beta = 0.05$, which requires $B_c = 9.0$ T, is not completely unreasonable, but is certainly only marginally possible with today's field technology. That observation, together with the fact that we would operate at the margin of stability to achieve $\beta = 0.05$, makes an ordinary Tokamak a questionable candidate for a reactor. The bottom line, however, is that a relatively modest increase in β to ~ 0.1 could make a Tokamak a much stronger reactor possibility.[j] As we will discuss later, just such an increase in β may be possible by using Tokamaks with noncircular cross sections.

We now investigate how our two MHD results limit the range of plasma densities and currents over which stable Tokamak operation can be achieved. Let us first note that from Ampere's law, we can find B_p at the plasma surface, in terms of the total current I, to be

$$B_p(a) = \frac{\mu_0 I}{2\pi a}$$

so that

$$q(a) = \frac{2\pi B_t}{\mu_0 R} \frac{a^2}{I}$$

Thus, $q(a) \gtrsim 2.5$ clearly specifies a maximum plasma current allowable for stability:

$$I \lesssim \frac{2\pi B_t}{q(a)\mu_0}\left(\frac{a}{R}\right)^2 R \tag{3.7}$$

If we choose $a/R = 1/3$ in a reactor, use $R = 7.5$ m from our earlier design calculation, and take $B_t \approx B_c = 9.0$ T from the preceding discussion,[k] then we must require I to be less than about 15 MA.

Experiments show that when q drops too low, the so-called "disruptive instability" occurs in which there are sudden large disturbances of the magnetic field and an explosive expansion of the plasma column, usually followed by termination of the plasma current. Strangely enough, the disruptive instability determines a low current limit for stable operation as well as the high current one. In a typical Tokamak discharge it is found that the plasma radius a shrinks rapidly as I decreases. Since $q(a) \propto a^2/I$, the decrease in a^2 can be faster than the decrease in I so that q drops below the critical value. The disruptive instability also limits the maximum plasma density in a Tokamak. For if we limit I, we limit β_p, which together with the requirement $\beta_p < R/a$ limits the plasma kinetic pressure $p =$

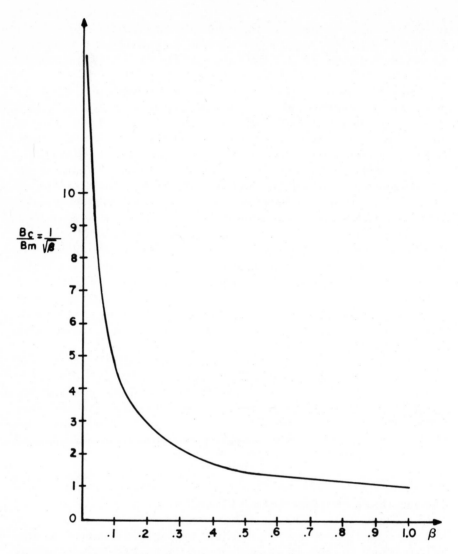

Figure 3–4. Normalized Required Confining Magnetic Field as a Function of β.

$nk(T_e + T_i)$ in a Tokamak with a given R/a. But if the temperatures are fixed by other considerations (heating rate, optimization of design, etc.), then a maximum density for stability is determined. This maximum density is easily seen to be given by

$$n < n_{\text{max}} \approx \frac{a}{6R} \frac{1}{k(T_e + T_i)} \frac{B_c^2}{2\mu_0} \tag{3.8}$$

(which is identical to the earlier requirement $\beta \lesssim a/6R$). In a reactor, with $a/R = 1/3$, $B_c = 9$ T, and $T_e = T_i = 10$ keV, we find $n_{max} \approx 5 \times 10^{20} m^{-3}$, which allows densities in the range of those considered in the reactor design example.

Figure 3–5 shows a sketch of the stable operating regime in the n-I plane, which summarizes the preceding discussion.[l] An additional feature shown in this figure is a low density boundary of the stable operating region in Tokamaks. This limit apparently occurs because at low enough densities, the velocity distribution function becomes sufficiently anisotropic, and hence non-Maxwellian, to drive plasma (velocity space- or micro-) instabilities. The electron velocity distribution function becomes increasingly anisotropic as the plasma density decreases for two reasons. First, as the electron density decreases, the directed velocity of the individual electrons must increase to carry a given current I. This increasing drift velocity, of course, tends to make the velocity distribution function more and more asymmetric (less and less Maxwellian) and hence unstable. Collisions, on the other hand, are a stabilizing factor since they tend to randomize the velocities and therefore make the velocity distribution function symmetrical. Thus, as the density decreases, the stabilizing effects due to collisions are ultimately overcome by the destabilizing effect of the increasing drift velocity.[m]

Note that in the type of instability just discussed, the plasma does not behave as a simple fluid which can be characterized by its particle density, its temperature, and a fluid velocity. Rather, the details of the velocity distribution function played an important role in describing the plasma behavior. There are actually many such particle kinetic effects which can cause instabilities in a Tokamak plasma that is MHD stable. Some of these effects are described in the next section.

The Particle Kinetic Properties of Tokamaks

So far, we have mainly examined the fluid or MHD properties of plasmas in Tokamaks.[n] Much of the plasma behavior that may be important in Tokamak reactors, however, depends on particle kinetic effects, that is, on the details of the velocity distribution function.

In order to investigate these effects, we need to consider the possible trajectories of a single particle traveling in the fields of a Tokamak more carefully than we have before. Recall from our earlier discussion of toroidal devices with rotational transforms that the field lines spiral around the torus and that the particles more or less follow them. The advertised benefit of the spiraling field lines was that they caused the particles to spiral around the torus and thus cause the vertical drift motion of the particles to average to zero (provided that they travel around the torus fast

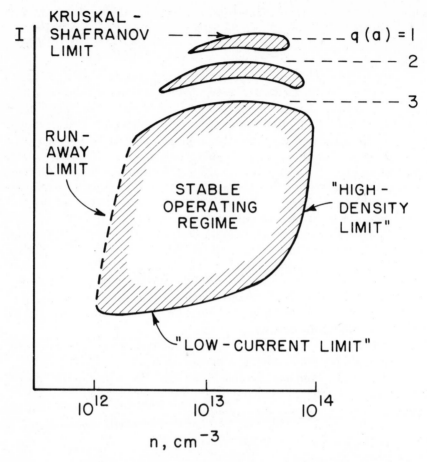

Figure 3–5. Rough Schematic of the Stable Operating Regime in Tokamaks. Disruptive instability sets in at both high and low currents and at high density. Electron runaway sets in at low
density (after Ref. [20]).

enough). The vertical drift we were trying to get rid of, remember, was
caused by the toroidal magnetic field being stronger near the inside of the
torus than near the outside. This inhomogeneity of the toroidal field has
another important effect. For as the particles spiral around the torus, they
see a region of strong magnetic field near the inside of the torus and a region of weaker field near the outside. But this means that as a particle
spirals around the torus, it sees what are effectively a sequence of magnetic mirrors. Some particles with low velocity along the field lines will be

reflected from these mirrors and then cannot spiral around the torus. The obvious question at this point is what happens to the reflected particles. Probably the first answer that comes to mind is that since these particles cannot spiral around the torus, their vertical drift will not be cancelled and they will be lost. The right answer turns out to be not quite as bad as that, as we will see later. Before we try to find the right answer to what happens to the particles that are reflected from mirrors in the rather complicated Tokamak fields, however, it is helpful to consider in more detail the behavior of particles reflected from mirrors in the simpler linear fields.

Single Particles in Magnetic Mirrors

Recall from the discussion of Figure 1–3 that a single particle spiraling around a magnetic field line experiences a $q\mathbf{v} \times \mathbf{B}$ retarding force due to its orbital velocity as it moves into a magnetic mirror region. What we did not point out during that discussion is that the velocity along the magnetic field lines together with the radial magnetic field component $\mathbf{B}_r$ gives the particle a $q\mathbf{v} \times \mathbf{B}$ kick in the azimuthal direction so that the trajectory of the particle forms a closed loop, as shown in Figure 3–6. Energy and angular momentum are, of course, conserved along the particle trajectory. Conservation of angular momentum can be put in a particularly convenient form as follows. First, break the particle velocity $\mathbf{v}$ into a component along the magnetic field $v_{\parallel}$, and into two components perpendicular to the field, $v_{\perp 1}$ and $v_{\perp 2}$. Roughly, $\mathbf{v}_{\perp}$ is the orbital motion around the magnetic field line. Next, let us calculate the angular momentum L in a region of constant magnetic field (as in the central portion of Figure 3–6 (a)). Now in such a region, $L = m v_{\perp} \rho_L$ where ρ_L, the Larmor radius, is the radius of the circular orbit around the magnetic field line. We can write that $v_{\perp} = \rho_L \omega_c$, where $\omega_c = q B_0 / m$, the cyclotron frequency at which the particle orbits the field lines, and B_0 is the magnetic field in the constant region. Thus, the angular momentum L of a particle in a constant magnetic field is

$$L_0 = \frac{m v_{\perp}^2}{\omega_c} = \frac{2m}{q} \frac{m v_{\perp}^2}{2B_0} = \frac{2m}{q} \frac{W_{\perp_0}}{B_0} \qquad (3.9)$$

where we have identified $W_{\perp_0} = \frac{1}{2} m v_{\perp}^2$ as the kinetic energy of the particle due to motion perpendicular to the field lines in the constant field region. What happens when the particle moves into the region where the magnetic field is not constant? Well, if the spatial variation of the magnetic field is slow enough and is $v_{\parallel}$ is small enough, then the particle continues to describe essentially circular orbits about the field lines at the local cyclotron frequency, $\omega_c = q B / m$ (B is the local value of the magnetic field). More specifically, the condition required is that B changes by only a small

B = B₀

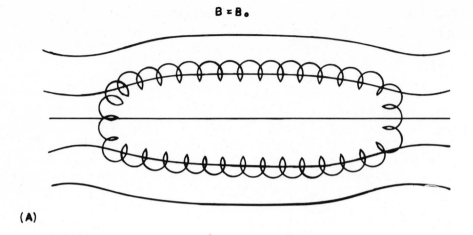

(A)

B = Bmin

B = Bmax   B = Bmax

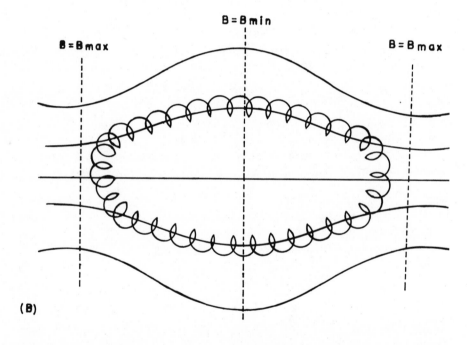

(B)

Figure 3–6. Trajectory of a Single Particle in Linear Magnetic Mirrors. (A) With Region of Constant Magnetic Field Separating the Magnetic Mirrors; (B) With No Region of Constant Magnetic Field Separating the Magnetic Mirrors.

fraction during an orbit. If this condition is met (usually easily done in practice) then, proceeding as before, we find the angular momentum at an arbitrary point in the orbit to be given by

$$L = \frac{2m}{q} \frac{W_\perp}{B}$$

Since $2m/q$ is just some number, we see that conservation of angular momentum requires the quantity $\mu = W_\perp/B$ (known as the magnetic moment) to be essentially constant along the trajectory.[o] Note in particular that μ is constant independent of the shape of the trajectory. That is, μ is constant along trajectories like the one shown in Figure 3–6 (b) and along wilder ones which we will see a little later.

We can use the invariance of the magnetic moment μ to investigate in more detail something we discussed earlier: particles with large enough $v_\parallel$ are not stopped (reflected) by a magnetic mirror. Consider Figure 3–6(a). The energy E of the particle, a constant, can be written as

$$E = \frac{1}{2}mv_\parallel^2 + \mu B$$

where $\mu = \frac{1}{2} mv_\perp^2/B$ is a constant, as we have just discussed. Since E is a constant, note that as the particle moves into a region of higher magnetic field, the velocity of the particle along the field $v_\parallel$ must decrease. If B is large enough, $v_\parallel$ goes to zero and the particle is reflected. The value of E can be found from

$$E = \frac{1}{2} mv_{\parallel 0}^2 + \mu B$$

where $v_{\parallel 0}$ is the value of $v_\parallel$ at the point where $B = B_{min}$. All particles that have an energy E which is less than μB_{max} clearly stop (are reflected) before reaching (and hence passing through) the point where $B = B_{max}$. Thus, the condition for a particle to be trapped within the magnetic mirror is that

$$E = \frac{1}{2}mv_{\parallel 0}^2 + \mu B_{min} < \mu B_{max} \qquad (3.10)$$

Rearranging this expression and using the fact that we can write $\mu = \frac{1}{2}mv_{\perp 0}^2/B_{min}$, where $v_{\perp 0}$ is the value of $v_\perp$ at the points where $B = B_{min}$, we obtain

$$\frac{v_{\parallel 0}}{v_{\perp 0}} < \sqrt{\frac{B_{max}}{B_{min}} - 1} \qquad (3.11)$$

for the trapping condition. This condition is sketched in Figure 3–7. If we define $\alpha \equiv \tan^{-1}(v_{\perp_0}/v_{\parallel_0})$ as the pitch angle of the orbit (where $B = B_{min}$), it is easy to see that the trapping condition can be written[p] as $\alpha > \alpha_c$ where

$$\alpha_c = \sin^{-1}\sqrt{\frac{B_{min}}{B_{max}}}$$

Many Particles in Magnetic Mirrors

If we consider a group of particles with a random distribution of velocities in a magnetic mirror field, we expect that some will escape and some will be trapped. To estimate the fraction trapped (for times less than the collision time), we can assume that the velocities are initially distributed isotropically with respect to the magnetic field. If each particle is represented as a point in velocity space (Figure 3–7), this means that the density f of points in velocity space is spherically symmetric and hence depends only on the radial coordinate $v = (v_\parallel^2 + v_\perp^2)^{1/2}$, the speed. Thus, $f = f(v)$. The fraction f_T of trapped particles is therefore

$$f_T = \frac{\text{number of trapped particles}}{\text{number of particles}} = \frac{\int_0^\infty v\,dv \int_{\alpha_c}^{\pi-\alpha_c} \sin\alpha\,d\alpha \int_0^{2\pi} d\phi f(v)}{\int_0^\infty v\,dv \int_0^{\pi} \sin\alpha\,d\alpha \int_0^{2\pi} d\phi f(v)}$$

$$(3.12)$$

where v, α, ϕ are respectively the radial, polar, and azimuthal coordinates in the spherical coordinate system in velocity space. Interchanging the order of integrations and cancelling some factors, we obtain

$$f_T = \frac{-\cos\alpha \Big|_{\alpha_c}^{\pi-\alpha_c}}{-\cos\alpha \Big|_0^{\pi}} = \cos\alpha_c = \frac{\sqrt{B_{max} - B_{min}}}{\sqrt{B_{max}}}$$

or

$$f_T = \sqrt{1 - \frac{B_{min}}{B_{max}}} \qquad (3.13)$$

Notice that the fraction of trapped particles increases as B_{max} increases.

Single Particles in Tokamak Fields

We have already noted that as the field lines in a Tokamak spiral around

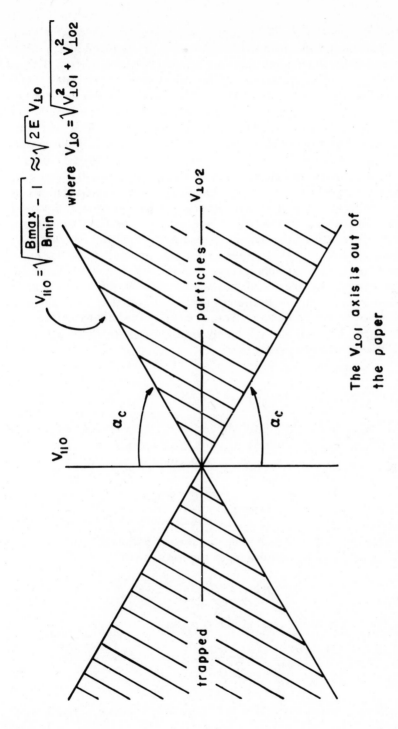

Figure 3–7. Velocity Space Regions for Trapped and Untrapped Particles. (The naught subscript indicates midplane quantities.)

the torus, they pass through what amounts to a sequence of magnetic mirrors because the toroidal magnetic field is stronger near the inside of the torus. Some of the plasma particles are trapped by these mirrors. Others remain untrapped. The projections of the trajectories of both types of particles onto a plane of constant ϕ are shown in Figure 3–8. For more or less obvious reasons, the trapped particle orbits are called "bananas." The bananas are roughly the warped counterparts of the trajectories[a] of Figure 3–6. The importance of the trajectories in Figure 3–8 is in examining what happens to a particle when it suffers a collision. As we discussed earlier in chapter 1, a collision can cause a particle to jump across the confining magnetic field. From Figure 3–8, it is clear that the maximum distance an untrapped particle can jump as a result of a single collision is about twice the Larmor radius, $2\rho_L$. A trapped particle, on the other hand, can jump as far as Δr_T, the width of the banana. Since Δr_T is typically considerably larger than ρ_L, it is clear that trapped particles can leak out of the Tokamak much faster than untrapped ones. In particular, the width Δr_T of the banana turns out to be $\Delta r_T \approx 2q\rho_L/\sqrt{\epsilon}$ where $\epsilon \equiv a/R$. With $q \approx 2.5$ and $\epsilon = 1/3$, then $\Delta r_T/\rho_L \approx 8$. For future reference, we note that the frequency, ω_b, with which a particle bounces from one end of the banana to the other and back again is given approximately by $\omega_b \approx v_t\sqrt{2\epsilon}/Rq$, where $v_t = (2kT/m)^{1/2}$ is a thermal velocity.

Many Particles in Tokamak Fields

At this point, it seems natural to wonder whether the fraction of trapped particles is large enough to make this enhanced leakage rate a potentially serious problem. Well, recall from the preceding section that the fraction of trapped particles is

$$f_T = \sqrt{1 - \frac{B_{\min}}{B_{\max}}}$$

Furthermore, since $B \approx B_t$ (since $B_p << B_t$) and B_t falls off as the inverse of the distance from the major axis of the torus, then

$$B \propto \frac{1}{R + r \cos \theta}$$

If the minor radius of the plasma is $r = a$, then

$$\frac{B_{\min}}{B_{\max}} = \frac{\dfrac{1}{R + a}}{\dfrac{1}{R - a}} = \frac{1 - \epsilon}{1 + \epsilon} \tag{3.14}$$

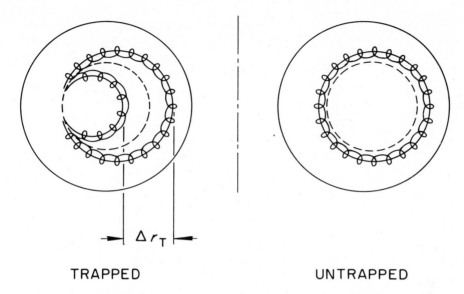

Δr_T

TRAPPED   UNTRAPPED

Figure 3–8. Projections of Particle Trajectories for Trapped and Untrapped Particles onto a Plane of Constant ϕ (after Ref. [20]).

where $\epsilon = a/R$ is the inverse aspect ratio of the Tokamak. Thus, the fraction of trapped particles is

$$f_T = \sqrt{\frac{2\epsilon}{1 + \epsilon}} \approx \sqrt{2\epsilon} \qquad (3.15)$$

Since in a reactor $\epsilon \approx 1/3$, we might expect $f_T \sim 0.8$, a large fraction. Thus
the radial diffusion of trapped particles across the confining magnetic field
can be expected to be an important process.

In order to investigate the loss of particles, both trapped and untrapped, quantitatively, we take time out to develop a simple model for
the process, which is useful not only for Tokamaks, but for discussing
other devices that we treat later, as well.

*A Random Walk Diffusion Model for Cross-Field
Particle Loss*

Consider Figure 3–9, which shows a typical plasma density profile in a
plane perpendicular to the confining magnetic field. (In a Tokamak, the
plane would be one of constant ϕ and the transverse coordinate would be
r). In particular, we consider a region between x and $x + \delta x$, which is nar-

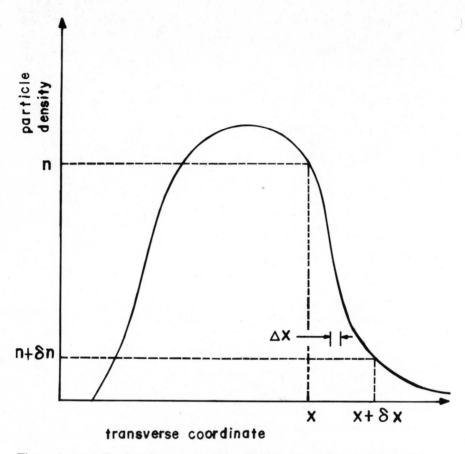

Figure 3–9. A Typical Plasma Density Profile in a Plane Perpendicular to the Confining Magnetic Field.

row enough that the fall-off in density is approximately linear. We assume that the particles jump a mean distance Δx in a random direction as a result of each collision, or what not, and that the average time between jumps is Δt. If we assume that $\Delta x \ll \delta x$, then it makes sense to model the process as a random walk (see Appendix 3–A). From symmetry, there is a net movement of particles only along the x direction from, say, x to $x + \delta x$. In that case, we have $\delta x = \sqrt{N}\Delta x$ where N is the average number of steps required to move a distance δx in steps of Δx. Let the time required to move a distance δx be δt. Since we are using a random walk model, then of course $\delta t = N\Delta t$. We can therefore write

$$\delta x = \sqrt{\frac{\delta t}{\Delta t}} \Delta x$$

and define some sort of diffusion velocity, v_D, as

$$v_D \equiv \frac{\delta x}{\delta t} = \frac{\Delta x}{\sqrt{\delta t \Delta t}}$$

The flux of particles (number of particles per m² per sec) to the right (actually in both directions) from x is $n v_D$. The flux of particles to the left (again, actually in both directions) from $x + \delta x$ is $(n + \delta n) v_D$. The net flux Γ of particles which move across the region between x and $x + \delta x$ is

$$\Gamma = [n - (n + \delta n)]v_D = \frac{-\delta n \Delta x}{\sqrt{\delta t \Delta t}} \tag{3.16}$$

or

$$\Gamma = -\frac{\delta n}{\delta x} \frac{(\Delta x)^2}{\Delta t} \tag{3.17}$$

Since by hypothesis δx is small enough for $n(x)$ to be considered linear between x and $x + \delta x$, then we have

$$\Gamma = -D \frac{dn}{dx} \tag{3.18}$$

where $D = (\Delta x)^2 / \Delta t$ is the diffusion coefficient for the process. Note that the diffusion, and hence the loss of particles, is driven by the density gradient. The strength of the diffusion, however, is determined by the mean step size, Δx, and the mean time between steps, Δt, for the particular process of interest.

Diffusion Due to Collisions in Tokamaks

In applying the random walk model to particle diffusion across a magnetic field, we need to know both the mean time between particle collisions and the average size of the jump caused by the collisions. Let us first consider the classical case of particles in a constant, uniform, magnetic field. From an earlier discussion, we recall that the trajectory of a particle is changed primarily by the cumulative effect of small random deflections by relatively distant particles rather than by the occasional close approach to another particle. It is therefore hard to say just when a "collision" has occurred. A common criterion, however, is to say that a collision has occurred when the particle trajectory has changed (by means of particle interactions) 90° with respect to its initial direction of motion so that the particle has lost all of its momentum along the original direction.[r] The frequency with which such collisions occur, denoted $\nu_{90°}$, can be calculated and turns out to be different for electrons and ions. In either case, we take the mean time between collisions to be $\Delta t = 1/\nu_{90°}$.

The next question is what to use for the step size, Δx. If we refer to Figure 1–5 and imagine a particle deflection of 90° rather than 180°, we see that the appropriate step size for a 90° collision is something like the Larmor radius, ρ_L. Thus, the coefficient for classical diffusion is $D \sim \nu_{90°}\rho_L^2$. Notice that $D \sim 1/B^2$ so that particle losses decrease rapidly with increasing magnetic field, a point we discussed earlier.[s]

Let us now consider particle diffusion in the more complicated toroidal geometry for the collisionless case in which particle trajectories are the banana orbits of Figure 3–8. As we have already mentioned, the banana width, Δr_T, is the relevant step size. To choose the collision frequency $1/\Delta t$ appropriate for a step size of Δr_T, consider Figures 3–7 and 3–8. In order to jump a distance Δr_T, the particle must reverse its direction of motion along the magnetic field line—that is the sign of $v_{\parallel_0}$ must change. For a typical trapped particle, Figure 3–7 shows that this sign change requires a change in pitch angle of something like $\sqrt{2\epsilon}$, since the maximum change in pitch angle for a trapped particle is $\sim 2\sqrt{2\epsilon}$. Thus, for Δt, we need to use the time it takes for small angle Coulomb scattering to accumulate to an angle of $\sqrt{2\epsilon}$. This time will generally be somewhat smaller than $1/\nu_{90°}$, but we can relate them as follows. Recall from our earlier random walk discussion of small angle Coulomb scattering that the net amount of deflection increases as the square root of the elapsed time. Using this fact and the fact that the angular deflection associated with $\nu_{90°}$ is $\pi/2$ while the angular deflection for the present case of interest is $\sqrt{2\epsilon}$, we can write

$$\frac{1/\nu_{\text{eff}}}{1/\nu_{90°}} = \frac{(\sqrt{2\epsilon})^2}{(\pi/2)^2} \tag{3.19}$$

Thus, roughly,

$$\nu_{\text{eff}} \sim \frac{\nu_{90°}}{\epsilon} \tag{3.20}$$

where ν_{eff}, the reciprocal of the time required for an angular deflection of $\sqrt{2\epsilon}$, is the effective collision frequency. Note that toroidal effects make the effective collision frequency, ν_{eff}, for the trapped particles several times greater than the collision frequency $\nu_{90°}$, in the classical case of particles in a linear magnetic field. The radial flux of trapped particles is

$$\Gamma = -\nu_{\text{eff}}(\Delta r_T)^2 f_T \frac{dn}{dx} = -D \frac{dn}{dx} \tag{3.21}$$

where we have included a factor f_T (fraction of particles trapped) to account for the fact that the density gradient of the trapped particles is $f_T dn/dx$, n being the combined density of trapped and untrapped particles. Recalling that the banana width is

$$\Delta r_T \approx \frac{2q\rho_L}{\sqrt{\epsilon}}$$

we see that the diffusion coefficient for this case is

$$D \approx \nu_{\text{eff}}(\Delta r_T)^2 f_T \approx 4\sqrt{2}\,\nu_{90°}\rho_L^2 q^2\epsilon^{-3/2} \tag{3.22}$$

This coefficient, which takes into account toroidal effects, is the so-called neoclassical diffusion coefficient. It is appropriate when the trapped particles are collisionless in the sense that their effective collision frequency ν_{eff} is less than the trapped particle bounce frequency ω_b, so that they can complete a banana orbit before suffering a collision. Note that the neoclassical diffusion coefficient is larger than the classical one by a factor something like

$$4\sqrt{2}\,q^2\epsilon^{-3/2} > 100$$

for $q \approx 2.5$ and $\epsilon \approx 1/3$. Thus toroidal effects trap certain of the plasma particles and, paradoxically, make it possible (through relatively large banana orbits and an enhanced collision frequency) for these particles to diffuse across the magnetic field and escape at a much more rapid rate than the untrapped particles. (In the context of the present discussion, the untrapped particles would be expected to diffuse across the torus at roughly the classical rate since the mean step size and collision frequency for these particles should not be changed much by bending the field lines into a torus.) Since the trapped particles form an appreciable fraction ($\sqrt{2\epsilon}$) of the plasma particles, the particle loss across a toroidal field can be expected to be many times that in a linear field, other things being equal. (The big advantage of the toroidal confinement field is, of course, that there are no end losses.) Nevertheless, even these enhanced neoclassical loss rates are quite low enough to make it fairly easy to confine a thermonuclear plasma.[t] For example, we know from earlier discussions[m] that $\nu_{\text{eff}} \propto 1/v_{th}^3$ while $\omega_c\rho_L = v_{\text{orbital}} \sim v_{th}$ so that

$$D \sim \rho_L^2 \nu_{\text{eff}} \sim \frac{1}{v_{th}} \sim T^{-1/2} \tag{3.23}$$

where ω_c is the cyclotron frequency of the particles as they gyrate about the magnetic field lines and $v_{th} \sim \sqrt{kT/m}$ is a characteristic thermal velocity. Thus, as the temperature in Tokamaks increases to thermonuclear values, neoclassical theory tells us that particle losses due to collisions should drop nicely. Unfortunately, neoclassical diffusion rates are not observed in Tokamak experiments. Rather, a somewhat higher, but, still relatively low, rate of diffusion, called anomalous diffusion, thought to be due to plasma instabilities, is observed. There is concern that as the temperature is increased so that the collision frequency decreases, then other instabilities which may be collisionally overdamped in present Tokamaks (which are relatively cold) will become underdamped, grow, and cause even more enhanced or anomalous diffusion. Thus, the neoclassical theory should be viewed as giving a lower bound to particle diffusion in Tokamaks.

Anomalous Diffusion in Tokamaks

Recall from an earlier discussion that microinstabilities with very short wavelengths are almost impossible to stabilize because shear in the magnetic field suppresses only instabilities whose wavelength is long enough for them to notice that the magnetic field is changing direction. Considering a given amount of shear and considering the large number of modes of instability in a plasma, we expect there to be short wavelength instabilities present in Tokamaks. These instabilities are driven by the free energy associated with the plasma having pressure (density and/or temperature) gradients at its edges because it is confined, or by free energy related to the plasma not having a Maxwellian velocity distribution function.

In any case, as an instability grows, it produces a kind of fine grained turbulence in which there are regions of both very strong and relatively weak electric fields. As we mentioned earlier, the particles scatter off the regions of large electric field in much the same way that they do off each other. The effective collision frequency is therefore increased and enhanced diffusion results.

It is very difficult to calculate the details of this process because the relevant equations are extremely nonlinear. We can, however, estimate the diffusion coefficient in a fairly simple way. Recall from our random walk discussions of diffusion that $D = (\Delta x)^2/\Delta t$ where Δx is the jump size and Δt is the time between jumps. For our case of diffusion across the magnetic field, we choose the scale size or the mean distance between adjacent regions of localized electric field (measured in the direction across the magnetic field) as Δx. This in turn should be something like $\lambda_\perp$, the wavelength measured perpendicular to the magnetic field for the instability at small amplitudes. The quantity $\lambda_\perp$ can be calculated by linearizing the relevant equations for simplicity. We expect the time between jumps to be a few (amplitude) e-folding times of the linearized instability because after only a few e-folding times, the electric fields should be strong enough to bump the particle. Again, from the linear theory, we find that as long as the amplitude of the instability is small, it increases as $e^{\gamma t}$ where γ is the growth rate. The e-folding time is thus $1/\gamma$ and we choose $\Delta t \sim 1/\gamma$. We thus conclude that the diffusion coefficient due to instabilities is

$$D \sim \gamma/k_\perp^2 \tag{3.24}$$

where $k_\perp = 2\pi/\lambda_\perp$ and we neglect numerical factors of a few. This result, though crude, is very important in that it permits us to estimate the answer (probably an upper bound in fact) to a highly nonlinear problem from the answers to a much simpler linear problem.

Let us now take a look at some microinstabilities which may be important in limiting the operation of Tokamak reactors. Although instabilities due to the velocity distribution function not being an isotropic Maxwellian[u] might be present in Tokamak reactors, it is generally believed that the most important microinstabilities will be driven by boundary effects (e.g., pressure gradients—gradients in n and/or T), which are present in any confined plasma.[v] As an example, consider Figure 3–10 which shows ripples in the edge of the plasma as a wave propagates the short way (θ-direction) around the torus.[w] At a particular point on the plasma surface, the particles oscillate in the radial direction as the wave passes by. From the Lorentz force law, $\mathbf{F} = q\mathbf{v} \times \mathbf{B}$, oscillation of the plasma, together with the toroidal magnetic field $\mathbf{B}_t$, induces poloidal velocities $v_{e\theta}$ and $v_{i\theta}$ for the electrons and ions respectively. (Note that $v_{e\theta}$ and $v_{i\theta}$ will have different signs.) The instantaneous radial force acting on the ions and electrons in a small volume element of the plasma is

$$F_r = (q_i v_{i\theta} + q_e v_{e\theta})B_t \tag{3.25}$$

The work done on this small volume element is proportional to

$$F_r\Delta r = (q_i v_{i\theta}\Delta r + q_e v_{e\theta}\Delta r)B_t \tag{3.26}$$

where B_t is a constant. In the simplest case, $v_{i\theta} \propto v_r$ and $v_{e\theta} \propto v_r$ where v_r is the radial velocity of the plasma volume element due to the wave. In particular, $v_{i\theta}$ and $v_{e\theta}$ are in phase with v_r. Note that since the oscillations are sinusoidal, v_r and Δr and hence $v_{i\theta}$ and Δr and $v_{e\theta}$ and Δr, are 90° out of phase. In the simplest case, therefore, the time average of $v_{i\theta}\Delta r$ and $v_{e\theta}\Delta r$ vanish so that $\langle F_r\Delta r\rangle_{av} = 0$. Thus, there is no net work done on the plasma. Suppose, however, that, because of collisions or whatever, $v_{i\theta}$ and/or $v_{e\theta}$ get out of phase with v_r and hence are no longer precisely out of phase with Δr.[x] Then $\langle F_r\Delta r\rangle_{av} \neq 0$. Specifically, if $v_{i\theta}$ or $v_{e\theta}$ begins to lag Δr in phase then $\langle F_r\Delta r\rangle_{av} > 0$ which means that the average radial force is directed so that it increases Δr, and hence the size of the wave. This, in turn, increases v_r and hence $v_{i\theta}$ and $v_{e\theta}$ which increases F_r, and so forth. The phase lag of $v_{i\theta}$ and/or $v_{e\theta}$ has thus resulted in a growing wave or an instability.

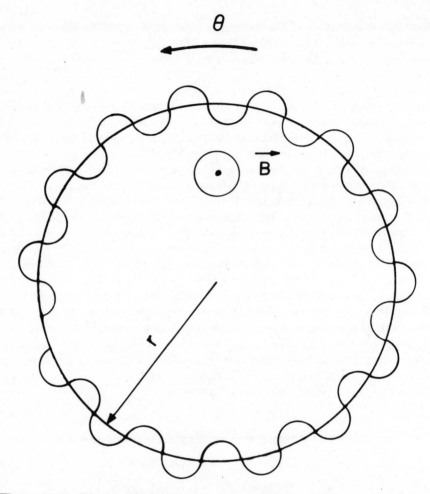

Figure 3–10. Ripples (Drift Waves) on the Plasma Surface Propagating the Short Way (θ-Direction) around a Torus.

It might seem that phase lags in $v_{e\theta}$ or $v_{i\theta}$ due to collisions should not be much of a worry in reactor plasmas since they are very hot and hence relatively collisionless. In fact, however, reactor plasmas will be just about hot enough so that the electrons and the ions (barely) are in the banana region in which a trapped particle is able to complete a banana orbit before experiencing a collison. We have already seen that the trapped particles have a much higher effective collision frequency than untrapped particles. This point, coupled with the fact that the trapped particles are relatively localized and hence are less affected by the stabilizing effects of shear, helical fields and the like means that the trapped particles offer an efficient route for the free energy of a confined plasma to drive instabilities which let the plasma escape across the magnetic field. Trapped particle modes or instabilities in which the relevant phase lags are caused by collisions are called resistive or dissipative drift waves, while those in which the phase lags are caused by something else (some sort of resonance effect, for example) are called collisionless drift waves.[y]

Although the theory of trapped particle modes is presently crude, the answers it provides give warning that as Tokamaks get hotter and more collisionless, these modes may cause significant or even disastrous increases in particle transport. Consider, for example, the case in which both the electrons and the ions are in the banana regime. This regime is the one expected in an operating reactor. Application of $D \sim \gamma/k_\perp^2$ gives [20]

$$D \sim \left(\frac{k}{2e}\right)^2 \frac{\epsilon^{5/2}}{\nu_{ei}} \left(\frac{T}{B}\right)^2 \left(\frac{1}{n}\frac{dn}{dr}\right)^2 \tag{3.27}$$

where ν_{ei} is $\nu_{90°}$ for electron-ion collisions. The relevant instability is a dissipative trapped ion mode. Note that although D does decrease as $1/B^2$ even in this regime, it increases rapidly with increasing T. This result does offer some hope, however, of decreasing D by controlling dn/dr by adjusting the density profile. With the boundary diffuse (dn/dr small) rather than sharp, the drift waves of Figure 3–10 tend to be suppressed.

Diffusion, Confinement Time, and Scaling in Tokamaks

Consider a small volume element of plasma in which there is a particle density n, a total number of particles N, and in which particles flow out with a flux Γ (particles/area × sec), where Γ in general varies over the volume element. We have

$$\frac{\partial N}{\partial t} = \int_V \frac{\partial n}{\partial t} dV = -\int_s \Gamma \cdot \mathbf{a}_n \, dS = -\int_V \nabla \cdot \Gamma \, dV \tag{3.28}$$

where S is the surface around V, $\mathbf{a}_n$ is the normal to S, and we have used Stokes's Theorem. Since the two volume integrals must be equal for arbitrarily shaped volumes, it must be true that

$$\frac{\partial n}{\partial t} + \nabla \cdot \Gamma = 0 \tag{3.29}$$

This equation, called the continuity equation, is just an expression of the conservation of particles. From our discussion of diffusion, we know that the particle flux is in the direction opposite to the gradient of the particle density:

$$\mathbf{\Gamma} = -D\mathbf{\nabla} n$$

The continuity equation therefore becomes the diffusion equation.

$$\frac{\partial n}{\partial t} - D\nabla^2 n = 0$$

To apply this equation to cross-field diffusion in a Tokamak, we consider a short sector of the torus and assume approximate cylindrical symmetry. We also neglect poloidal variations in density. In addition, we use separation of variables to find that if the plasma production takes place at $t < 0$ and stops at $t = 0$, then the plasma density decays exponentially in time as it diffuses out of the torus: $n(r, t) = n_0(r)e^{-t/\tau}$, where τ, the e-folding time, is taken as the confinement time. With these assumptions and results, the diffusion equation simplifies to Bessel's equation:

$$\frac{d^2 n_0(r)}{dr^2} + \frac{1}{r}\frac{dn_0(r)}{dr} + \frac{1}{D\tau}n_0(r) = 0$$

If we apply the boundary condition that $n_0(0)$ is finite, then we find the solution is

$$n_0(r) \propto J_0\left(\frac{r}{\sqrt{D\tau}}\right)$$

where J_0 is the zero order Bessel function. If we further require $n_0(a) = 0$, so that the density is zero at the edge of the plasma, then

$$J_0\left(\frac{a}{\sqrt{D\tau}}\right) = 0$$

The first root[z] of $J_0(x) = 0$ occurs at $x = 2.4$. Thus, we find[aa]

$$\tau = \frac{a^2}{5.76D} \tag{3.30}$$

Notice that the confinement time increases as the plasma radius a is increased. This result is the main motivation for building larger machines—the bigger they are, the better the confinement should be. A simple interpretation is that it takes longer for a particle to random walk out of a larger machine. Recall, however, that the important quantity in plasma confinement for thermonuclear purposes is not τ, but $n\tau$. In Tokamaks, the plasma density n is limited by the condition $\beta_p \lesssim R/a$, which is a requirement for equilibrium. In particular, we must have

$$n \lesssim \frac{R}{a} \frac{B_p^2}{2\mu_0 kT} = \frac{R}{a} \frac{\mu_0 I^2}{8\pi^2 a^2 kT}$$

since $B_p \approx \mu_0 I/2\pi a$. Thus the upper bound on $n\tau$ for a Tokamak is

$$n\tau \approx \frac{\mu_0 I^2}{5.76(2\pi)^2 2\epsilon DkT} \tag{3.31}$$

If we assume that the relevant D is the one associated with the dissipative trapped ion mode, then

$$n\tau \approx \frac{\pi \ln \Lambda}{5.76\sqrt{m_e}}\left(\frac{e^3}{16\pi^3}\frac{\mu_0}{\epsilon_0}\right)^2 \epsilon^{-9/2} \frac{B_t^2}{(kT)^{11/2}} I^4 \tag{3.32}$$

where $\ln \Lambda \approx 20$. We have used the fact that [25]

$$\nu_{ei} = \frac{1}{4\pi\epsilon_0^2} \frac{e^4 n}{\sqrt{m_e}(3kT)^{3/2}} \ln \Lambda$$

and have used the upper limit on density which results from $\beta_p \lesssim R/a$. For simplicity, we have also set the characteristic length for density variation across the cross section, $[(1/ndn/dr)]^{-1}$ equal to the minor radius a. If we assume $\epsilon = 1/3$, $B_t = 5T$ and $T = 13$ keV, then we find that the current required to achieve $n\tau = 10^{20}$ sec m^{-3} would be about 5 MA. Perhaps 10 MA would be appropriate for a reactor. If we require $q = 3$ in such a reactor, then $q = \epsilon B_t/B_p$ and $B_p \approx \mu_0 I/2\pi a$ require $a = 3.6$ m and $R \approx 11$ m. These estimates are crude and probably optimistic since the effects of impurities, which are certain to be present, have been neglected. Nevertheless, it is encouraging to notice that the size of the Tokamak determined here is comparable to the size of the toroidal reactor we arrived at earlier by a completely different approach.

Tokamaks with Noncircular Cross Sections

We saw in an earlier section that Tokamaks with circular cross sections are limited to βs of about five percent. Operation at such low βs, we saw, is really not economically attractive in a reactor. We also mentioned, however, that Tokamaks with noncircular cross sections offer the potential of operating at the higher βs needed for a reactor. We now investigate this point.

Consider a Tokamak with one of the noncircular cross sections shown in Figure 3–11. On a magnetic surface of such a device, recall that the pitch angle γ of the magnetic field is such that $\tan \gamma = B_p/B_t$. As before, $B = (B_t^2 + B_p^2)^{1/2}$ so that

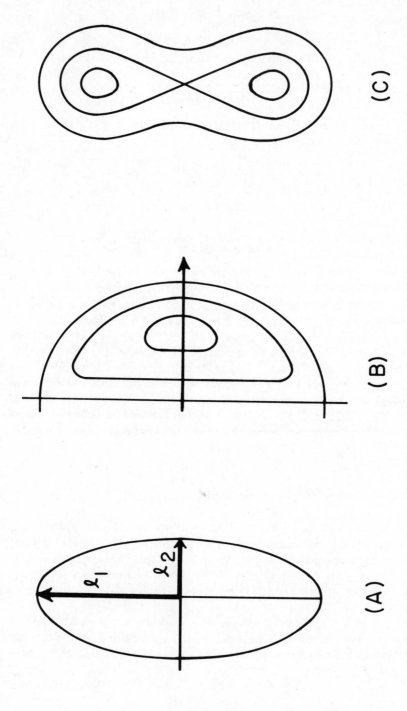

Figure 3–11. Examples of Tokamaks with Noncircular Minor Cross Sections: (A) Ellipse, (B) D-Shape, (C) Doublet. The major axis of the torus is on the left (after Ref. [20]).

$$\beta = \beta_p / [1 + (B_t/B_p)^2]$$

where $\beta_p = 4\mu_0 nkT/B_p^2$ is the poloidal beta for $T_e = T_i = T$. Now the poloidal distance traversed by a field line of pitch γ as it passes once around the torus in the toroidal direction is $2\pi R \tan \gamma = 2\pi R B_p/B_t$ as we found before. But in terms of the rotational transform ι this same distance is about $\iota \ell / 2\pi$ where ℓ is the circumference of the cross section. (For a circular cross section Tokamak with minor radius a, for example, $\ell = 2\pi a$.) Thus $B_t/B_p = (2\pi)^2 R/\iota\ell$, so that

$$\beta = \frac{\beta_p}{1 + \left(q\dfrac{2\pi R}{\ell}\right)^2}$$

since $q \equiv 2\pi/\iota$.

At this point, we recall that β_p was limited to R/a for the circular case and wonder what the corresponding limit might be for the noncircular cross-section Tokamaks. The argument in our earlier discussion, remember, was that as a straight magnetic field is bent into a torus, the resulting magnetic field gradients make confinement by the toroidal field less effective. It therefore needs to be supplemented by confinement from an additional poloidal field, which must support a fraction of at least a/R of the kinetic pressure.[bb] From physical considerations, it seems that we should interpret a as a measure of how far the plasma extends across the toroidal field gradient (and hence of the plasma width) rather than as a measure of the height of the plasma. The fraction a/R therefore seems to be the ratio of the plasma half width to the radius of curvature of the magnetic field. With this interpretation, it would seem that the result $\beta_p \lesssim R/a$ holds for noncircular as well as circular, Tokamaks. More detailed considerations bear out this conclusion. Thus

$$\beta \lesssim \frac{R/a}{1 + \left(q\dfrac{2\pi R}{\ell}\right)^2} \approx \frac{R}{a}\left(\frac{\ell}{2\pi Rq}\right)^2$$

Notice that if we try to increase β by adjusting the plasma cross section, we need to keep the plasma width, $2a$, small and increase its minor perimeter, ℓ. These conditions suggest the use of elongated cross sections such as those shown in Figure 3–11. The ellipse of Figure 3–11(a) turns out to be unsatisfactory since detailed calculations show that the safety factor q must be raised to such an extent to ensure stability that the effects of the increased plasma perimeter are canceled out. This result follows from the rather sharp curvature of the magnetic field lines toward the plasma at the top and the bottom of the ellipse. Plasma instabilities (MHD type) tend to develop in such regions of sharp curvature of the field lines toward the plasma because this means that the magnetic field is decreasing in strength as one moves away from the plasma (Appendix 1–A). In such regions, therefore, the plasma, rather than being in a magnetic well, or minimum-B configuration, is on a sort of magnetic hill. Such a state corresponds, of course, to an unstable equilibrium.

The D shape of Figure 3–11(b) turns out to be better than the ellipse primarily because it concentrates more of the plasma near the inside of the torus where the toroidal field lines bend away from the plasma (favorable curvature).

A Tokamak with the cross section of Figure 3–11(c) is called a doublet. Note that this configuration contains a closed separatrix of the magnetic field inside the plasma in the form of a figure 8. The net result of having such a separatrix inside the plasma is the presence of a large shear in the magnetic field. This shear allows $\ell \lesssim 2\pi R$ while maintaining stability at a fairly low value of $q(\sim 3)$.

At the time of writing, computer studies are underway to determine the optimum plasma cross section for a Tokamak. It presently appears, however, that stability considerations limit the elongation of the cross section to a degree such that [26] $\beta_{max} \approx 0.1$—an increase of perhaps a factor of two over a standard circular cross-section Tokamaks.

Tokamaks in the Future

In the Magnetic Confinement Systems Program at ERDA, Tokamaks are presently considered to be the most promising approach to achieving commercial fusion power and have been given development priority.

Two large Tokamaks are scheduled to begin operation in 1978. Doublet III, shown in Figure 3–12 is designed to study confinement and plasma transport in the various trapped particle regimes typical of reactor conditions in a device with a noncircular cross section. The full nominal design current is 5 MA for the design value of $q(a) = 2.6$. If the scaling is governed by trapped particle microinstabilities, as predicted, then the 5 MA current alone should result in 2.5-keV ion temperatures at particle densities of 10^{14} cm^{-3}. The planned 10 MW of injected neutral beam power should then result in a plasma with 5-keV ion temperature at a density of 2

This section consists largely of more or less verbatim extracts from *Fusion Power Research and Development: Summary Report on Magnetic Confinement Experiments, Report ERDA 76–34*, January 1976 (prepared by the Confinement Systems Program, Division of Controlled Thermonuclear Research, U.S. Energy Research and Development Administration—for sale by the U.S. Government Printing Office) and the first draft (June 4, 1976) of Volume III of the *Five Year Plan for Fusion Power Research and Development Program* (prepared by the Division of Magnetic Fusion Energy, U.S. Energy Research and Development Administration).

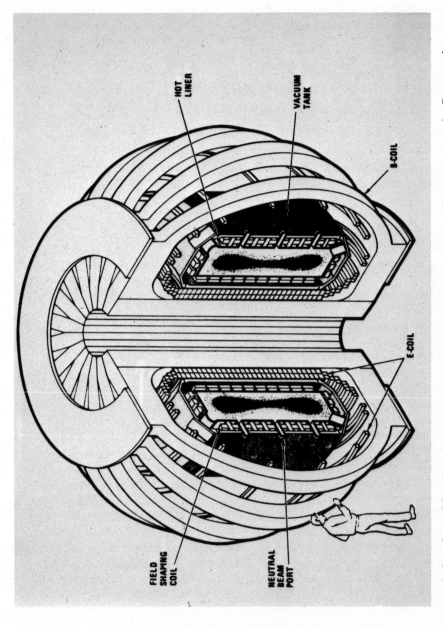

Figure 3–12. A drawing of Doublet III (courtesy of USERDA) to be built at General Atomic Corporation.

Table 3–1
Parameters for Doublet III

A. Machine Parameters for Doublet III

	$B_t(T)$	$a(m)$	$R(m)$	R/a	q	$I(kA)$
Maximum Design	2.6	0.45 × 1.50	1.40	—	2.6	5000

B. Typical[a] Parameters for Doublet III

$n(m^{-3})$	$T_e(keV)$	$T_i(keV)$	$\tau(msec)$	$n\tau(m^{-3}\text{-sec})$	β_p
2 × 10²⁰	4	4	250	0.5 × 10²⁰	2

[a] Assuming trapped ion scaling in a doublet configuration, 2.6 T magnetic field and 12 MW of auxiliary heating.

Table 3–2
Parameters for PDX

A. Machine Parameters for PDX

	$B_t(T)$	$a(m)$	$R(m)$	R/a	q	$I(kA)$
Maximum Design	2.4	0.47	1.45	3.1	3.6	500

B. Typical Plasma Parameters for PDX

$n(m^{-3})$	$T_e(keV)$	$T_i(keV)$	$\tau_e(msec)$	$n\tau_e(m^{-3}\text{-sec})$	β_p
2 × 10¹⁹	1	<1	100	~2 × 10¹⁸	~1

$\times 10^{14}$ cm^{-3}. The coils are designed for $4T$ operation in the event that upgrading is desirable. The Doublet III parameters are shown in Table 3–1. The estimated fabrication cost of Doublet III is 27.7×10^6.

The second large Tokamak, which is scheduled to come on line in 1978, is the Poloidal Divertor Experiment (PDX), shown in Figure 3–13, whose parameters are given in Table 3–2. The primary objective of PDX, which will be located in the Princeton University Plasma Physics Laboratory, is to evaluate the use of additional poloidal field coils to divert the edge of the plasma to a separate collection chamber before the ions interact with the wall. Such a divertor will reduce the level of plasma impurities sputtered from the wall. In addition, it will define the plasma size and shape without requiring the mechanical limiters used in many present Tokamaks. For successful operation, PDX should provide:

(1) Substantial reduction in impurity level
(2) Detailed information on the behavior of a plasma with a divertor at its edge, and on any deleterious effects introduced by the divertor

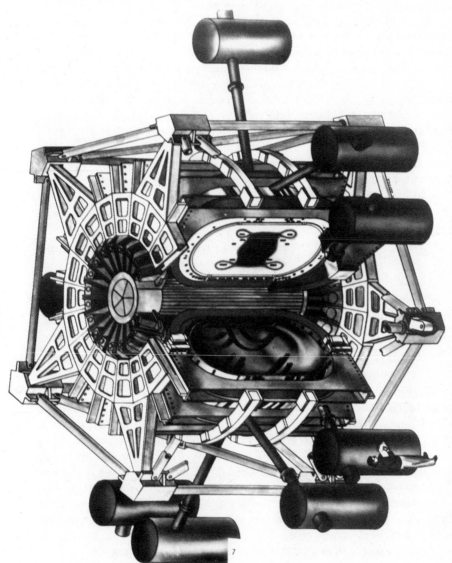

Figure 3–13. The Poloidal Divertor Experiment (PDX) (courtesy USERDA).

(3) An adequate technological data base for the design of divertors for larger Tokamaks and reactors

The estimated fabrication cost of PDX is now 18.8×10^6.

After Doublet III and PDX, the next step for Tokamaks is the first D-T burning Tokamak, the Tokamak Fusion Test Reactor (TFTR), which is scheduled to begin operation in 1981. The TFTR will be the nation's first magnetic confinement fusion device to demonstrate experimentally the release of fusion energy from the deuterium-tritium reaction under conditions projected for future experimental power reactors. It will represent an intermediate step between relatively small zero-power physics experiments, such as Doublet III and PDX, and future experimental reactors planned for the mid-1980s. The TFTR will be located at the Princeton University Plasma Physics Laboratory near Princeton, New Jersey. The construction project is scheduled to be completed in 1981 at a total cost of 228×10^6.

The TFTR has major objectives in both physics and engineering. The principal objectives are:

(1) To demonstrate fusion energy production from the burning of D-T in a magnetically confined toroidal plasma system

(2) To build a neutral beam heated Tokamak in which hydrogen, deuterium, and D-T plasma can be generated in order to:

(a) Study the physics of large Tokamaks

(b) Verify advanced engineering concepts for D-T Tokamak systems

(3) To demonstrate experimentally that sufficient physics and engineering understanding of large fusion systems exists in advance of construction of an experimental power reactor

The TFTR is a two ion component device (TCT) in which an energetic beam of ions travels through a warm target plasma.[cc] The advantage of such a scheme is that a much greater burn of D-T fuel can be achieved at a given $n\tau$ and temperature T. In TFTR, the two ion component approach may make it possible to reach energy break-even (although this is not listed as an objective) in the Lawson sense with $n\tau$ and T lower than the values we mentioned in chapter 1, which were based on an isotropic Maxwellian velocity distribution for the ions. The two ion component approach is not likely to be used in a commercial reactor because of the large amount of (costly) circulating power required. It may be useful, however, in obtaining reactor-like conditions in a smaller experiment (lower $n\tau$) and hence in an experiment which will be more quickly available.

The specific objectives of the TFTR project are:

(1) Attain reasonably pure hydrogenic plasma conditions at 5–10 keV

temperature, approximately $10^{20}m^{-3}$ density, and provide stable confinement with $n\tau_E$ equal to or greater than $10^{19}m^{-3}$ sec.[dd]

(2) Provide a neutral beam injection system capable of injecting 20 MW of a 120-keV D^0 beam into the plasma for at least 0.5 sec.

(3) Provide a toroidal magnetic field of about 5 T (on the vacuum chamber axis), for at least a 3-sec flat-top time, with a 5-min repetition rate.

(4) Develop plasma handling techniques and provide hardware capable of initiation, control (including feedback control and major radius compression) and dissipation of Tokamak discharges up to 2.5 MA.

(5) Provide a vacuum chamber of adequate size (2.7 m major radius and 1.1 m minor radius), equipped for high-power discharge cleaning and capable of achieving base pressures below 5×10^{-8} Torr.

(6) Provide capability for routine pulsed operation with H-H, D-D, D-He^3, or D-T plasmas, with safe and reliable gas handling and support systems.

The principal design features of TFTR are:

Size

2.7-m major radius

Plasma

2.5-MA maximum plasma current
1.7-m diameter
Target plasma conditions: $n\tau \approx 10^{19}m^{-3}$sec; $T_i \approx 3$–4 keV

Neutral Beam Injection

120–150 keV

Magnets

Water-cooled copper
5.0-T center line field
Provision for heating by adiabatic compression

Power Supply

4500-MJ, 660-MW pulsed
AC motor/generator/flywheel

Special Features

Tritrium supply and containment system
Shielded magnets and auxiliary systems
Remote handling capability

Energy Release

1–10 MJ with D-T

Beyond upgrading TFTR (as well as other large Tokamaks), primarily by adding more auxiliary heating power, the plans for future Tokamaks depend both upon technical progress and funding. It is presently assumed, however, that the next step for Tokamaks will be either a Prototype Experimental Power Reactor or an Ignition Test Reactor (PEPR/ITR) which will be built in the mid-1980s. At about this same time, an engineering test reactor (FERF/ETR), which could be a Tokamak, is also expected to be built. By the early 1990s, an Experimental Power Reactor (EPR) which would make net electrical power with high reliability would be built. This device would be followed by the Fusion Power Demonstration Reactor in about 1998.

Vigorous programs to develop fusion reactors, primarily of the Tokamak type, also exist in the Soviet Union, Europe, and Japan. The Soviets are planning a D-T burning Tokamak, the T-20, which is much larger than the TFTR. T-20 is currently planned for operation in 1983. The Euratom[ee] countries have formed a joint fusion development program, which is completing the design of a large experiment called the Joint European Tokamak (JET). The Japanese are planning a TFTR-sized Tokamak (JT-60) which, however, will not burn D-T. It is scheduled to operate in 1980. Appendix 3–B gives a comparative list of parameters for these machines, the TFTR and preliminary experimental power reactor designs by Argonne National Laboratory (ANL EPR), General Atomics (GA EPR), and Oak Ridge National Laboratory (ORNL EPR).

Notes

(a) Tokamak is a kind of Russian acronym for "toroidal chamber magnetic."

(b) The heating rate decreases because the effective collision frequency for momentum transfer drops as the temperature increases. See discussion in a later note.

(c) This result follows from Ampere's law at radius r:

$$I_{\text{enclosed}} = \frac{1}{\mu_0} \oint \mathbf{B} \cdot d\mathbf{l} \approx 2\pi r B_p / \mu_0$$

(d) The use of a transformer of course limits the Tokamak to pulsed operation. Conceptual design studies indicate that the maximum pulse length might range from a few minutes to perhaps an hour. There has been speculation that a so-called "bootstrap" toroidal plasma current [20], which might occur as a result of unavoidable diffusion processes in the Tokamak, might permit steady state operation of a Tokamak reactor by taking the place of the transformer-induced toroidal plasma current. At present, this prospect seems like a long shot. For further discussion see George H. Miley, *Fusion Energy Conversion* pp. 213–215, 430–435 [22].

(e) The triplet (r, θ, ϕ) forms an orthogonal set of coordinates known as toroidal coordinates. Unfortunately for the theory of toroidal devices, most of the partial differential equations encountered in plasma physics are not separable in toroidal coordinates (even though they are separable in some other orthogonal systems such as cartesian, cylindrical, spherical, etc.). This circumstance makes the analysis of toroidal devices more difficult than that of linear ones.

(f) Recall from our earlier discussion of the Kruskal-Shafranov limit that stability resulted when the charged particles could gain access to an entire magnetic surface (and hence neutralize any charge accumulations on the surface) simply by flowing freely along magnetic field lines. Our simple argument did not take into account, however, that collisions impede the free flow of particles along the field lines. When collisions and the radial increase in r/B_p are considered, theory indicates that $q > 1$ must be replaced by the more stringent requirement $q(a) \gtrsim 2.5$ for stable operation. Actually, theory seems to indicate that there are, in addition, some small ranges of $q(a)$ between 1 and 3 which might permit stable operation (see Figure 3–5). Because of the small sizes of these regions, however, they are usually thought to be unimportant in practice.

(g) The existence of an equilibrium and its MHD stability are clearly two different questions. In this particular case, however, the existence condition and the MHD stability condition turn out to be roughly the same.

(h) The physical basis of this requirement is discussed from a different point of view in the later section in this chapter on Tokamaks with noncircular cross sections.

(i) For Tokamaks with noncircular cross sections, β_{max} might possibly be somewhat larger, as we will discuss later.

(j) For the D-D reaction, recall that $B_{min} = 6.4$ T. Thus, even with $\beta \approx 0.1$, we must have $B_c \approx 20$ T, a seemingly impossibly large magnetic field in the volume required. It seems clear, therefore, that Tokamaks, and other low β devices are reactor possibilities only for the D-T reaction. The use of high β ($\beta > 0.1$) devices appears to be essential for any future reactor that uses D-D, D-He^3, etc., reactions.

(k) With $R/a > 3$, then $q(a) \gtrsim 2.5$ requires $B_t \gtrsim 7.5 B_p$. Thus, the confining field $B_c \approx B_t$.

(l) The two small regions of stability at the top of the figure are the ones we discussed in an earlier note as being too small, and hence unimportant.

(m) The situation is actually a little more complicated than that just described in that the anisotropy of the velocity distribution function is caused mainly by so-called "runaway" electrons, usually a small fraction of the electrons in the plasma. The runaway electron phenomenon is based on the fact that the collision frequency (for momentum transfer) of an electron in a plasma actually decreases as the electron moves faster. This circumstance means that as the electrons are accelerated to carry current, they become less likely to be slowed down by collisions. Some few of the electrons can therefore reach a state of very high energy (perhaps a few hundred keV). These electrons are the runaway electrons. Their velocity is limited primarily by radiation damping due to the production of hard x-rays.

The fact that the effective collision frequency (for momentum transfer) in a plasma actually decreases as the electron moves faster seems, at first, unreasonable. After all, it is certainly clear that a fast-moving particle encounters more particles per second than a slow moving one. In a plasma, however, the vast majority of these "encounters" merely lead to a small random deflection of the moving particle as it passes by (but not very near) the other plasma particles. (In typical thermonuclear plasmas, the particle density is low enough that a "binary collision," in the sense of a single encounter producing a big deflection of the particle trajectory, happens only rarely.) Thus, by a collision of an electron in a plasma, we usually mean an appreciable deflection of its trajectory due to a series of small random deflections. The particle therefore undergoes a random walk away from the direction of its original trajectory. There are two competing factors at work, however. First of all, the individual deflections are larger for a slow particle because it is subject to the deflecting force of the nearby charged particle for a longer period of time. More specifically, if the mean time between the individual encounters is Δt and the average acceleration on the particle during Δt is α, then the transverse deflection of the trajectory during a single encounter can be found by integrating the equation $\ddot{r} = \alpha$ to obtain

$$\Delta r \sim \alpha(\Delta t)^2$$

On the other hand, a fast particle has more encounters with scattering centers during a given time. In order to determine whether fast or slow particles are deflected more in the long run, suppose that during a time τ an electron has N encounters with scattering centers (ions) so that $\tau = N\Delta t$. Since the electron takes a random walk in the transverse direction, the probable transverse displacement after N encounters, $r(\tau)$, is $\sqrt{N}$ times the step size, Δr (see Appendix 3–A). That is

$$r(\tau) \sim \sqrt{N}\alpha(\Delta t)^2$$

or

$$r(\tau) \sim \alpha\sqrt{\tau}\,(\Delta t)^{3/2}$$

But

$$\Delta t \approx n^{-1/3}v$$

where $n^{-1/3}$ is the average spacing between scattering centers (ions) in a plasma of density n, and v is the electron velocity. Thus,

$$r(\tau) \sim \alpha\left(\frac{\tau}{n}\right)^{1/2}\frac{1}{v^{3/2}}$$

Thus, fast particles are deflected much less than slow ones during a given time and hence can be said to "collide" with the plasma less frequently. By way of interpretation, note that the fact that a fast particle undergoes more deflections than a slow one is dominated by the fact that the size of the deflections for a fast particle are very much smaller.

(n) An exception, of course, was the velocity space instability that leads to the lower limit on density for stable operation of Tokamaks.

(o) Such a quantity, which is constant for sufficiently slow changes of some set of parameters, is called an *adiabatic invariant*.

(p) In mirror machines, the cone $\alpha < \alpha_c$ is called the loss cone (see Figure 3–7).

(q) The actual trajectories are not closed, but precess around the torus (see Ref. [21]). The projections are closed, however.

(r) Choosing ninety degrees seems plausible since it is the average angle of deflection per collision of randomly moving hard spheres scattering off fixed (or at least very massive) scattering centers, a case in which what is meant by a collision is very clear.

(s) Since $\nu_{90°}$ and ρ_L are different for ions and electrons, the diffusion coefficients are also different. When the particles start to diffuse at different rates, however, charge separation sets up strong electric fields, which have the net effect of locking the electrons and ions together so that they diffuse at an intermediate rate. This process is called ambipolar diffusion. Not surprisingly, the ambipolar diffusion coefficient also scales as $1/B^2$.

(t) This circumstance still holds, even for the more collisional cases in which the electrons or both the electrons and the ions undergo a collision, on the average, before completing a banana orbit (see Ref. [20]).

(u) The toroidal current is carried by the untrapped particles, which thus have an appreciable toroidal velocity. This makes the velocity distribution anisotropic and non-Maxwellian.

(v) These instabilities are sometimes called universal instabilities. Many of them can be suppressed by shear, etc., but some of the short wavelength ones we discuss here may thrive in a reactor.

(w) For reasons not obvious from our discussion such a wave is called a drift wave.

(x) The ions and electrons tend to maintain identical radial velocities since different radial velocities result in charge separation and hence large restoring electric fields, as we have discussed before. A difference in poloidal velocities can occur more easily since it does not result in charge separation.

(y) The special roles trapped particles play in driving drift waves unstable, as well as other aspects of trapped particle instabilities, are described in W. Manheimer, *My First Trapped Particle Instability Reader*, USERDA, Washington, D.C. (1977).

(z) Using the first root amounts to assuming that the initial density profile, $n_0(r)$, is reasonably smooth (see Ref. [23]).

(aa) A somewhat different solution gives $\tau = a^2/4D$, a result that is often quoted (see Ref. [24]).

(bb) To consider this result from a different viewpoint, we note from Figure 3–3 that as we proceed radially outward from the minor axis of the torus, we reach a boundary between an inner region, in which the field lines spiral about the minor axis, and an outer region in which the field lines are basically vertical. This boundary is called a separatrix. Now recalling that B_v is supposed to counteract the outward expansion of the current carrying plasma, we have $B_v \propto I \propto B_p$ so that $B_v \propto B_p$. The constant of proportionality turns out to be [21] about $\beta_p a/R$. Thus

$$B_v \approx \left(\frac{a}{R}\beta_p\right)B_p$$

Now we have already seen that we would like to choose β_p as large as possible to maximize β. But increasing β_p increases B_v. When $\beta_p \approx R/a$, then $B_v \approx B_p$. At this point, B_v is clearly no longer negligible in determining the Tokamak field configuration. In fact, as B_v increases in comparison with B_p, the separatrix moves closer to the plasma boundary. For $B_v \approx B_p$ (that is, $\beta_p \approx R/a$), the separatrix, and hence the vertical field begins to invade the plasma. The plasma is then able to flow along the vertical field lines and escape. Thus, maintaining equilibrium in a Tokamak requires $\beta_p \lesssim R/a$.

(cc) See note (i) in chapter 1 and note (e) in chapter 2.

(dd) τ_E is the energy confinement time, which can be shorter than the particle containment time τ_p, if the more energetic particles escape faster than the less energetic ones.

(ee) Euratom is a cooperative nuclear research and development organization composed of Western European countries.

Appendix 3–A

The Random Walk

Consider a particle which moves around in a plane by sequentially jumping a distance Δr (the same on each jump) in an arbitrary direction (generally different on each jump). Thus, the change in the position of the particle on the ith jump can be written as

$$\Delta \mathbf{r}_i = \Delta r[\cos \phi_i \hat{x} + \sin \phi_i \hat{y}]$$

where ϕ_i indicates the direction of the ith jump. A natural question to ask is "Where is the particle after N jumps?" If we call its position after N jumps $\mathbf{r}(N)$, then we can write

$$\mathbf{r}(N) = \sum_{i=1}^{N} \Delta r[\cos \phi_i \hat{x} + \sin \phi_i \hat{y}]$$

$$\mathbf{r}(N) = \Delta r \sum_{i=1}^{N} [\cos \phi_i \hat{x} + \cos \phi_i \hat{y}]$$

This answer is fine except that we do not know the ϕ_i. We do know that they are randomly distributed between 0 and 2π however, so that for large enough N we might expect

$$\sum_{i=1}^{N} \cos \phi_i = \sum_{i=1}^{N} \sin \phi_i = 0$$

Thus, $\mathbf{r}_{av}(N) = 0$ for large N, which is a little disappointing since it seems very unlikely that after N jumps the particle is back where it started. If we give the matter a little thought, however, we see that since we have no idea in which direction the particle will have migrated after N jumps, the result $\mathbf{r}_{av}(N) = 0$ is the best we can do about predicting the position of the particle in an average sense. Since the uncertainty in direction seems to be the source of the trouble, suppose we try to answer the simpler question of how far, on the average, has the particle traveled after N jumps, without worrying about the direction of its travel. In terms of a polar coordinate system with its origin at the particle's starting point, we want to calculate the probable value for the radial coordinate, r, without worrying about the angular coordinate, θ. To accomplish this, we note that the square of the distance traveled after N jumps is

$$\mathbf{r}(N) \cdot \mathbf{r}(N) = \left\{\Delta r \sum_{i=1}^{N} [\cos \phi_i \hat{x} + \sin \phi_i \hat{y}]\right\} \cdot \left\{\Delta r \sum_{j=1}^{N} [\cos \phi_j \hat{x} + \sin \phi_j \hat{y}]\right\}$$

$$= (\Delta r)^2 \sum_{i=1}^{N} \sum_{j=1}^{N} [\cos \phi_i \cos \phi_j + \sin \phi_i \sin \phi_j]$$

$$= (\Delta r)^2 \sum_{i=1}^{N} \sum_{j=1}^{N} \cos (\phi_i - \phi_j)$$

All the terms with $i \neq j$ average to zero for large N. Since there are N terms with $i = j$, we get

$$\mathbf{r}(N) \cdot \mathbf{r}(N) \approx (\Delta r)^2 N$$

for large N, for the mean square of the distance traveled. The root mean square estimate of the distance traveled is therefore:

$$r = \sqrt{\mathbf{r}(N) \cdot \mathbf{r}(N)} \approx \sqrt{N} \Delta r$$

Appendix 3–B

Table 3–B–1
Large Tokamak Parameters List

Parameter	Symbol	Unit	JET	TCT(TFTR)	T-20	JT-60	ANL EPR	GA EPR	ORNL EPR
1. Major Radius	R	m	2.96	2.48	5	3	6.25	4	6.75
2. Plasma Minor Radius	a	m	1.25	0.85	2	2	2.1	1	2.25
3. Plasma Half Height	b	m	2.10	0.85	2	1	2.1	3	2.25
4. Aspect Ratio	R/a		2.37	2.9	2.5	3	3.0	3.1	3
5. Elongation Ratio	b/a		1.68	1.0	1	1	1	3	1
6. Magnetic Field on Axis	B_T	T	3.4	5.2	3.5	5	3.4	4	4.8
7. Maximum Field	B_{max}	T	6.9	9.5	7.8	11	7.5	8	11
8. Plasma Current	I	MA	4.8[a]	2.5	6	3.3	4.8	13.5	7.2
9. Safety Factor at Surface	q_a		6	3.0	2.3	3.5	2.5	2.5	2.5
10. Safety Factor at Axis	q_0		1	1.2	1.1	1	1.2	~1	~1
11. Current Rise Time		sec	1.0	0.05	1.1	0.1–1	1	1	~1–2
12. Mean Ion Temperature	T_i	keV	5	6.0	7–10	5–10	9.6	15	8
13. Mean Ion Density	$\bar{n}$	m^{-3}	5×10^{19}	8×10^{19}[d]	$(0.5–5) \times 10^{19}$	$(2–10) \times 10^{19}$	5.6×10^{19}	1.8×10^{20}	$\sim 7.5 \times 10^{19}$
14. Mean Electron Temperature	$\bar{T}_e$	keV	5	6.0	7–10	5–10	10	15	8
15. Plasma Pressure Ratio (Poloidal Beta)	β_p		1	1.0	1	1	2.2	2.2	1.9
16. Plasma Pressure Ratio (Toroidal Beta)	$\bar{\beta}_T$		0.03	0.007	0.03	0.02	~0.03	~0.14	~0.033
17. Energy Confinement Time	τ_E	sec	1	0.015[e]	2	0.2–1	7(10 × TIM)	3	6
18. $\bar{n}\tau_E$		m^{-3}-sec	0.5×10^{20}	1.5×10^{19}[d]	10^{20}	$(2–6) \times 10^{19}$	4×10^{20}	5×10^{20}	5×10^{20}
19. Pulse Length (Flat Top)		sec	20	1.0	5–20	10	~50	60	100
20. Interval between Pulses		sec	600	300	240	600	15	20	~0
21. Vessel Internal Radius		m	1.32 Horiz./ 4.18 Vert.	1.1	2.1	1.2	2.4	3.3/ 1.15 horiz.	2.25

Item	Units							
22. Type of Divertor		Limiter (?)	None	None	Magnetic Limiter Radially Outside	None	None	None
23. First Wall Surface Material		Inconel	AISI-305 SS	SS 0 × 18 HST[f]	Mo, SiC, C	SS/(C,SiC) liner	SS/(C,SiC) liner	SS/(C) liner
24. Wall (Blanket/Liner) Temperature during Operation	°C	20–50	50–80	100–500	500–Room Temp.	600/1500	650/1700	650/1400
25. Bake Out Temperature	°C	500	—	600	500	400	300	—[h]
26. Vessel Volume	m³	190	64	400	100	711	400	550
27. Vessel Surface Area	m²	450	110	400	1200 (net)	592	448	596
28. Pumping Speed	ℓ/sec	3×10^4	3×10^4	1×10^5	2×10^4	8×10^5	1.5×10^5	1.6×10^6
29. Base Pressure	torr	2×10^{-10}	4×10^{-8}	2×10^{-8}	10^{-9}	10^{-9}	10^{-8}	—
30. Neutral Injection Power	MW	3–25	40	60	10–20	40	200	100
31. Neutral Injection Energy	keV	80, 160	150; 120 & 60	80–160	50–100	180	—	200
32. Neutral Injection Pulse Length	sec	0.3–10	0.5; 0.1	2–13	10	3	1	~5
33. Number of Neutral Injection Lines (and sources)		6–24	6 (24)	8–(32)	24–(24)	16–32–(32)	—	6–(12)
34. RF Heating Power	MW	3–20	—	60	~10	—	—	—
35. RF Frequency	MHz	~800	—	~1000, ~60	~1000	—	—	—
36. RF Pulse Length	sec	—	—	2–13	~10	—	—	—
37. Number of Toroidal Field Coils		32	20	24	24	16	24	20
38. Shape of Toroidal Field Coils		D-shape	Circular	DM	Circular	Pure Tension	Pure Tension D	Min. Bending Oval
39. Internal Radius of Toroidal Field Coils	m	1.56(horiz)	1.4	2.7	1.94	3.9	3.3	3.7
40. MEG-AMP-Turns of Toroidal Field Coil	MAT	51	64.4	86.4	75	105	80	162

108

Table 3–B–1 (cont.)

Parameter	Symbol Unit	JET	TCT(TFTR)	T–20	JT–60	ANL EPR	GA EPR	ORNL EPR
41. Average Current Density for TF Coils	A/m²	25 Max	4×10^7	1.6×10^7	1.8×10^7 at center	1.28×10^7	1.33×10^7	1.78×10^7
42. Resistive (or Refrigeration) Power for TF Coils	MW	280	327	800	230	3.65	5.3	5.35
43. Peak Power for TF Coils	MW	330	448	1200	350	—	—	—
44. Toroidal Field Ripple at Plasma Edge	±%	±3.5	±0.5	±1	±0.05	+2	±0.5	±2.5
45. MEG-AMP-Turns in Ohmic Heating Coils	MAT	21	17	±20	±6.0	±45	175	45
46. Flux Swing in Core	V-sec	34	15	46	±15	±37	60	±65
47. Type of Core		Complete (Center Saturates)	Air	Air	Air	Air	Air	Air
48. Peak Power in Ohmic Heating Coils	MW	130[e]	34	600	?	~2000	~5000	~2000
49. MEG-AMP-Turns in Equilibrium Field Coils	MAT	3.2	4.5	12	2.25	10	10	~7
50. Peak Power in Equilibrium Field Coils		70[b]	200	420	?	90	54	16
51. Other Poloidal Windings		—	—	Feedback System	Mag. Limiter, Horiz. Field & Quad Field	—	—	shielding
52. Peak Power in Other Windings	MW	—	—	100	?	—	—	100
53. Peak Power Supply for Heating	MW	60	100	500	130	290	—	200(D⁻)
54. Total Peak Power	MW	~550	700	1680	?	~2000	~5000	~2000

	Units	2MG + FW, 2	AC Motor-Generator Flywheel	Grid, MG	Power Grid and 3 Motor Generators	Homopolar?	Power Grid Switching + MG Set	Power Grid Solid State Switching & Homopolar Generators
55. Energy Consumed per Pulse	M MJ	~10,000	4400	40,000	—	—	—	—
56. Type of Power Source		2MG + FW, 2	AC Motor-Generator Flywheel	Grid, MG	Power Grid and 3 Motor Generators	Homopolar?	Power Grid Switching + MG Set	Power Grid Solid State Switching & Homopolar Generators
57. Average Power Load	MVA	~25	40	130	100	<0	<0	<0
58. D-T Neutrons per Pulse		~10^{20}	10^{18}	10^{20}–10^{21}	No D-T	~2×10^{21}	~10^{22}	~4×10^{21}
59. 14-MeV Neutron Flux at Wall	n/m²- sec	2×10^{17}	4×10^{16}	10^{17}	"	~7×10^{16}	~5×10^{17}	~7×10^{16}
60. Expected Number of D-T Pulses during Life		1000–5000	4000	10^5	"	~5×10^5	~5×10^5	~5×10^5
61. Fluence at First Wall (14-MeV) (EPR, 1 yr.—50% duty cycle)	n/m²	(2–10) $\times 10^{23}$	5×10^{19}	8×10^{23}	"	~10^{24}	8×10^{24}	~10^{24}
62. Estimated Cost of Facility (Apparatus)		1.35×10^8 UC[c]	1.4×10^8 $US[e]	(3–4) $\times 10^8$ Rubles	—	500×10^6	500×10^6	500×10^6
63. Expected Date of Approval		Dec. 1975	1975	—	1976[g]	FY–78	FY–78	FY–78
64. Expected Date of Operation		1980	1980	—	1979[g]	1986	1986	1986

Appendix 3–B is taken from Appendix A of the *Summary of the Proceedings of the Workshop on Conceptual Design Studies of Experimental Power Reactor—I*, Michael Murphy and Jefferson Neff, Eds., Oak Ridge National Laboratory, Oak Ridge, Tennessee, September 9–10, 1975. (Available NTIS, as report No. ERDA–8A, of Division of Controlled Thermonuclear Research, USERDA, November 1975).

[a] JET ultimate performance: D-Shape.
[b] Taken as the peak power from grid or flywheel generator (excludes peaks from inductive storage).
[c] JET cost for basic performance only.
[d] In hot core region.
[e] In June 1975 dollars; includes buildings, 25% contingency; does not include engineering salaries.
[f] Changeable at the first step of experiment.
[g] Subject to government approval and funding.
[h] A dash indicates either the parameter is not applicable or the editors of ERDA–89 were unable to determine the value.

4

Mirror Machines

Up to this point, we have considered primarily magnetic confinement configurations (such as Tokamaks) which have no ends and are hence said to be closed. Although such configurations obviate the difficulties with end losses that plague magnetic mirror or open systems,[a] the resultant problems sometimes make one wonder if the medicine is not as bad as the malady. We are thus led to consider the potential of magnetic mirror devices for fusion reactors [1, 27].

Recall the simple magnetic mirror of Figure 1–3. Putting aside for a moment the problem of end losses, notice that there are regions just inside the mirrors in which the magnetic field lines are concave toward the plasma. As we mentioned in the previous chapter, such curvature means that the magnetic field is decreasing in strength away from the plasma so that the plasma in this region finds itself on a magnetic hill rather than in a magnetic well. To prevent the development of an MHD instability and the consequent loss across the magnetic field, therefore, we need to come up with a minimum-B (see chapter 1, "Magnetic Confinement") magnetic mirror configuration. One such configuration, produced by what is called for rather obvious reasons a baseball coil, is shown in Figure 4–1. Notice that the plasma cross section is more or less circular near the center and is flattened to orthogonal fan shapes in the mirror regions. Note also that the magnetic field lines are all convex toward the plasma so that the plasma sees an increasing B field in every direction. Thus, the plasma is confined in a minimum-B magnetic mirror field configuration. Figure 4–2 shows a plasma confined by a Yin-Yang coil system, which can be viewed as two slightly flattened baseball coils placed so that they produce a field configuration very similar to that produced by a baseball coil. The advantage of the Yin-Yang coil is that the resultant magnetic well is deeper and that the strength[b] of the mirrors can be controlled individually.

Now that we have a minimum-B magnetic field and hence have eliminated MHD instabilities, we turn to the problem of end losses. We know that particles that have high velocity along the magnetic field can pass through a magnetic mirror and escape. From the analysis of chapter 3, we know that the particles that can escape are those whose velocity lies within a certain cone in velocity space, called the loss cone. On the face of it, the loss cone does not seem to be much of a problem—we might simply adopt the strategy of throwing away the particles in the loss cone and

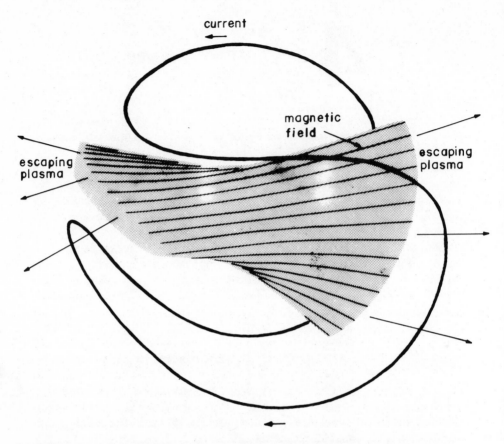

current

magnetic
field

escaping
plasma

escaping
plasma

Figure 4–1. Plasma Confined in a Minimum-*B* Magnetic Field Configuration Produced by a Baseball Coil (after Ref. [27]).

keeping the trapped particles for fusion. Such a strategy might be workable if it were not for particle collisions during which particles generally change their velocities and hence can end up in the loss cone even though they were trapped before the collision. Thus, all of the particles are eventually lost from the mirror region whether or not they are initially in the loss cone. Now, the rate of loss clearly depends on the collision frequency, which, we recall, decreases with temperature. From this point of view, therefore, we would like to choose the temperature in a mirror device to be as high as possible. There are two problems with this strategy. First of all, for a given particle density, very high temperatures mean very high kinetic pressure ($p = nk(T_e + T_i)$) which in turn could require very large magnetic fields. The thing that helps here is that there is no funda-

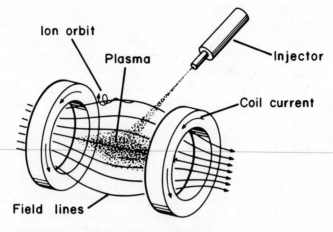

Simple Magnetic Mirror

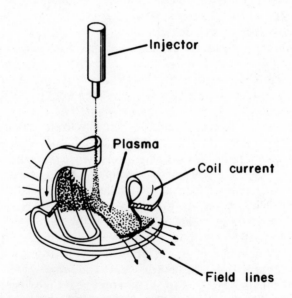

Minimum–B Magnetic Mirror
(Yin–Yang Coils)

Figure 4–2. Plasma Confinement in a Minimum-*B* Magnetic Mirror Produced by Yin-Yang Coils (Courtesy USERDA).

mental limit on β to ensure MHD stability in mirror devices as there was in Tokamaks (where necessarily $\beta \lesssim a/6R$). Thus, by letting $\beta \to 1$, we can contain a much hotter plasma with a magnetic field of a given strength in a mirror device than in a Tokamak. For example, since $\beta = 2\mu_0 nk(T_e + T_i)/B^2$, we see that for a given n and B, a device with $\beta = 0.9$ can support a plasma with the kinetic pressure twenty times higher than that which a device with $\beta = 0.045$ (a value typical of Tokamak reactor designs) can support.

The second problem with high temperature is radiation losses, which, as we have seen, increase rapidly with temperature. Of course we also saw that it is the electrons that radiate, but the ions that fuse. In principle, therefore, we would like to have T_e small and T_i large. In Tokamaks and most other fusion devices, the mean time between electron-ion collisions is short compared to the mean particle confinement time so that the electrons and ions are coupled very closely in a thermal sense. Thus T_e and T_i cannot be very different in such devices. In mirror devices, however, the relatively high operating temperatures lead to a decreased collision frequency (and hence to an increased mean time between electron-ion collisions) so that the ion temperature can be maintained at a value several times that of the electrons. Thus, we stand a chance of moderating radiation losses in mirror devices.

Although end losses caused by trapped particles being scattered into the loss cone by collisions can be dealt with by operating at high temperatures, we still have to worry about microinstabilities. These instabilities, recall, can lead to an effectively enhanced collision frequency, which of course is the very thing we do not want. The free energy for driving microinstabilities in mirrors comes primarily from velocity distribution functions that are not Maxwellian, although inhomogeneities in particle density and magnetic fields also play a role. (In Tokamaks, recall that the spatial inhomogeneities were the dominant driving force for microinstabilities.) The departure from a Maxwellian velocity distribution arises in two different ways. First, the loss cone means that there is a big hole in velocity space where particles ought to be if the velocity distribution function were Maxwellian. Second, the plasma ordinarily is maintained in a mirror device by injection of an energetic beam of neutral particles into a target plasma that ionizes the neutrals and randomizes their directed velocities so that they are trapped to sustain and heat the plasma (see Figure 4–2). The fact that the injected particles begin their life in the plasma as high energy ions therefore can lead to a bump at the high energy end of the velocity distribution—a definitely non-Maxwellian feature.

The distortion of the velocity distribution function (away from a Maxwellian) caused by beam injection can be reduced by appropriately adjusting the characteristics of the beam to make the bump more gentle (broader and not so high), for example. When all other tricks have been used up, however, the designer has one left which is not available to the designer of a closed system, such as a Tokamak: the length of the confinement region can be limited. Now why is this a useful trick? Well, recall that instabilities are simply growing waves. Furthermore, these waves ordinarily grow larger and larger as they propagate along through the plasma. In a certain sense, a closed confinement system, such as a Tokamak, looks like a plasma of infinite length to these waves as they propagate around and around the torus. In a mirror device, however, the waves can propagate for only a finite length,[c] and hence grow only a limited amount, which can be controlled by the length of the machine.

In summary, theory says that it should be possible to make minimum-B mirror devices stable if they are not too long. Even if the plasma is stable and we agree to operate at high temperatures to reduce the collision frequency, however, it turns out that we still have a problem of severe end losses.

Our first thought might be to make the device very long so that most of the plasma would be so far away from the ends that it hardly knows they are there. Of course, we just saw that this approach leads us down the primrose path as far as microinstabilities are concerned. Even if it did not, however, we probably would not insist on pursuing it since to make the device long enough to delay appreciably the escape of the very high energy ions used in a mirror machine would require a device at least several kilometers long. It seems, therefore, that we must learn to live with end losses in mirror devices.[d]

The problem with end losses is not so much the particle losses—replacing the particles with new ones by injection is not likely to be an insurmountable difficulty—rather, the problem is the *energy* these particles carry away. Thus, what we need to do to deal with end losses is to extract the energy from the escaping particles and re-inject it into the plasma. In extracting the energy, we have one important thing working for us: the energy is carried by charged particles so that it is possible to use high efficiency direct conversion techniques rather than low efficiency thermal cycles. The strategy, therefore, is to convert the kinetic energy of the escaping particles to electricity by means of a high efficiency direct conversion process and then to use the electrical energy to drive high efficiency neutral beam injectors, or perhaps rf heating, to recycle the energy into the plasma. The net result would be an effectively improved energy confinement time (limited by the conversion efficiencies) achieved at the cost of added complexity and capital cost.[e]

A sketch of one type of proposed direct conversion apparatus [22] is shown in Figure 4-3. Note that the apparatus consists of two main parts: an expander and a collector. In the expander region, the magnetic field

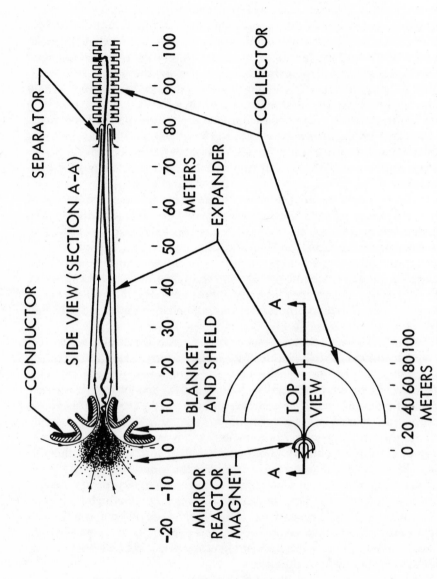

Figure 4–3. Apparatus for Converting End-Loss Ion Energy from a Mirror Device into DC Electrical Output (after Ref. [1]).

lines are spread out into a fan shape so that the B field decreases in strength. As we saw in chapter 3, the total energy E of a particle in a magnetic field of magnitude B can be written $E = \frac{1}{2}mv_\parallel^2 + \mu B$ where μ is a constant. Thus, as B decreases in the expander, the particle kinetic energy becomes mostly directed along the magnetic field. Since the particles are more or less tied to the magnetic field lines, the particles also fan out in the expanded region and the particle density is consequently decreased. At the end (edge) of the expander region, therefore, we have a relatively low density of particles streaming along the magnetic field lines. In fact, the expander is designed to dilute the particle density to such an extent that the particles no longer behave as a plasma, but behave as a collection of more or less independent charged particles whose dynamics are governed by the applied electric and magnetic fields rather than the fields from the surrounding particles.[f] Upon entering the collection region, therefore, the particles can be viewed as an incident beam of individual charged particles streaming along the magnetic field lines. Immediately upon the particles' entering the collection region, a curved magnetic field (relatively weak at this point) peels the electrons out of the beam and collects them on a single electrode. The ions are relatively unaffected by this magnetic field because of their much larger momentum together with the fact that the particle density is low enough that they are no longer strongly coupled to the electrons. Hence, the ions proceed further into the collection region. As they do so, they encounter a potential hill, which is maintained by applying appropriate potentials to the succession of individual electrodes that the ions encounter. As the ions climb the hill, they do work against the electric field which corresponds to the applied potential. When an ion has used up all of its kinetic energy in climbing the potential hill, it comes to rest and is collected on a nearby electrode as it starts back down the hill. Current will therefore flow through any load that might be connected between one of the ion collector electrodes and the single electron collector electrode. In a sense, this converter is an ion accelerator used in reverse. A disadvantage of this particular scheme is that each collector electrode produces current at a different voltage. To get a common voltage output from the convertor a combination of inverters and diodes and electric circuitry is required. The fewer ion collector electrodes there are, the simpler this circuitry is. The conversion efficiency, on the other hand, increases as the number of electrodes is increases so that no ion must travel backwards very far (from its stopping point on the potential hill) to be collected.[g] Obviously, some sort of economic compromise must be made in practice.

Notice that a mirror machine that uses direct conversion to recirculate the energy carried by escaping particles can be viewed as an energy or power amplifier in which the power absorbed by the plasma from injected neutral beams represents the input power, while the rate at which fusion energy is released (alphas and neutrons) represents the output power. Part of the output power of the reactor is fed back (positive feedback) to the input through the neutral beam injector, which is powered partly by the output of the direct converter and partly by electrical power from the main thermal conversion cycle. For operation at a steady power level, the fraction of output power that is fed back must be adjusted so that the loop power gain (taking into account the power gain of the reactor proper, the injector efficiency, the absorption efficiency of the neutral beam by the plasma, and the appropriately weighted efficiencies of the direct and thermal conversion cycles) must be unity. Break-even operation occurs when all of the output power must be used for recirculation with none left for external use. Economic operation of a reactor would probably require the power gain of the reactor proper to be high enough that only a small fraction of the output power would need to be recirculated to maintain the loop power gain at unity.[h]

To view the situation a little more quantitatively, consider Figure 4–4. Note that the reactor produces a thermonuclear power P_N, which appears partly as neutron kinetic energy and partly as kinetic energy of the alphas. The power input to the injector is P_I, which is transferred to the plasma with an efficiency η_I so that the power into the plasma is $\eta_I P_I$. The quantity

$$Q = \frac{\text{fusion power generated in the plasma}}{\text{power input to the plasma}} = \frac{P_N}{\eta_I P_I} \qquad (4.1)$$

is called the amplification factor.[i] If P_D is the power output of the direct converter and P_T is the power output of the thermal cycle, then the fraction ϵ of recirculating power is

$$\epsilon = \frac{P_I}{P_T + P_D} \qquad (4.2)$$

Now if h is the fraction of fusion power that appears as neutrons ($h = 0.8$ for D-T reactions), then we can write

$$P_T = h\eta_T P_N \qquad (4.3)$$

where η_T is the efficiency of the thermal cycle that converts the neutron kinetic energy into electrical power. The power available for direct conversion is the power released as alpha particles plus the power injected into the plasma: $(1 - h)P_N + \eta_I P_I$. If the direct conversion efficiency is η_D, then the power output of the converter is

$$P_D = \eta_D[(1 - h)P_N + \eta_I P_I] \qquad (4.4)$$

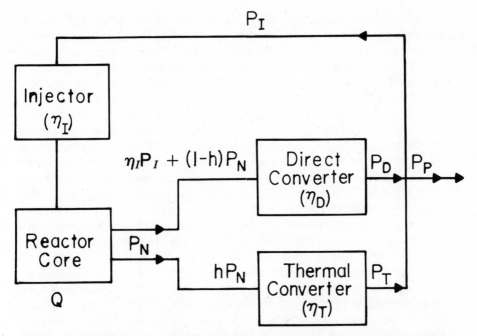

Figure 4–4. Power Flow Diagram Showing the Main Features of a Magnetic Mirror Fusion Power Plant with Direct Conversion (after Ref. [1]).

Substituting these results for P_T and P_D and using the fact that $P_I = P_N/Q\eta_I$, we find that

$$\epsilon^{-1} = \eta_I Q[h\eta_T + \eta_D(1 - h) + 1/Q] \qquad (4.5)$$

or, turning this result inside out,

$$Q = \frac{\epsilon^{-1} - \eta_D\eta_I}{\eta_I[h\eta_T + (1 - h)\eta_D]} \qquad (4.6)$$

To find the Q required for break even, we set $\epsilon = 1$ so that all of the output power is required just to sustain the reaction. If we consider the case with no direct conversion ($\eta_D = 0$) so that the energy of the escaping charged particles is simply abandoned, then the required Q is

$$Q = \frac{1}{\eta_I h\eta_T} = 3.125 \qquad (4.7)$$

if we choose, somewhat optimistically, $\eta_I = 0.8$ and $\eta_T = 0.5$. With a direct conversion efficiency η_D of 0.8, the required value is $Q = 1.08$. Thus, direct conversion can appreciably reduce the power amplification required in a device in order to achieve break-even operation. In a reactor, we would like for $\epsilon \lesssim 0.1$ to keep down the circulating power and, hence, the capital cost. In this case Q increases from 1.08 to 28.125. Unfortunately, detailed calculations show that even under ideal circumstances, it seems difficult to realize Q values larger than about one. Thus, it seems that mirror devices must operate with a large fraction of recirculating power.

Although we have spent the past few pages lamenting the existence of mirror losses as far as confinement is concerned, we recall the old saying, "It's an ill wind which blows no good," and note that there are positive aspects as well. For one thing, we have already seen that the end losses make direct conversion of the plasma thermal energy to electricity possible. This feature, important in the D-T cycle primarily as an energy recycling scheme, is potentially very important for reactions other than D-T in which most of the energy is given off in charged rather than neutral particles.[j] It is not obvious that a Tokamak, for example, is particularly suited for direct conversion even if it could be used for such reactions. Second, the end losses provide a built-in divertor, in that particles are automatically extracted from the system through the ends. This circumstance should reduce wall sputtering effects since most particles escape through the ends before striking the walls. Moreover, mirror losses turn out to be selectively high for high atomic number particles[k] and hence work to rid the plasma of impurities. Since the presence of plasma impurities is presently thought to be a major problem in reactor plasmas, this built-in divertor could be a significant advantage for mirror devices.

On the basis of dramatically improved performance[l] of mirror devices beginning in 1975 ($n\tau = 3 \times 10^{18}$ m^{-3}-sec, $T_i = 9$ keV) ERDA is considering funding the construction of a large mirror experiment, called MX, which is scheduled for operation in 1981. A limited number of D-T shots would be possible on this machine, which should come on line nearly concurrently with the Tokamak Fusion Test Reactor (TFTR). A major objective of this device will be to test confinement scaling for longer times to see if $n\tau$ scales as $T_i^{3/2}$ as classical theory predicts, and to test methods for improving the power balance between output and circulating power (increasing Q).[m] Assuming such an improvement is demonstrated, this device would be followed by a PEPR (Prototype Experimental Power Reactor) device in the late 1980s. The FERF/ETR mentioned earlier in the Tokamak section could also be a mirror. This device could be followed by an EPR operating in 1996 and a demonstration reactor around 2004. Table 4–1 lists machine parameters for the MX and a possible mirror FERF facility. The projected cost for the MX is $\$100 \times 10^6$.

Table 4–1
Machine Parameters for MX and a Projected Mirror FERF

	MX	FERF
$L_{\text{mirror-to-mirror}}$	3.4 m	4 m
B_{central}	2.0 T	3.7 T
R_m (vacuum mirror ratio)	2	2
Duration	continuous	continuous
Start-up Beam	1000 A, 20 keV	—
Sustaining Beams	750 A	500 A
	80 keV	65–100 keV
	10% duty cycle	continuous
$n\tau$	10^{18} m^{-3}-sec	1.5×10^{18} m^{-3}-sec
n	10^{20} m^{-3}	3×10^{20} m^{-3}
T_i	50 keV	70 keV

Adapted from F. H. Coensgen, "MX Major Project Proposal," LLL-Prop-142, Lawrence Livermore Laboratory, March 15, 1976.

Notes

(a) The bumpy torus, which we describe a little later, is a closed toroidal mirror system. Nevertheless, most mirror systems are open.

(b) The strength of the mirror is often measured in terms of the mirror ratio, which is the ratio of the B field in the mirror region to that at the center of the device. The mirror ratio should be greater than unity and usually the bigger the better, although the point of diminishing returns is actually reached with a mirror ratio of 2 or 3.

(c) For this to be true, it is necessary to smooth the spatial gradients (n, B, etc.) so that wave reflections are minimized, since a plasma of finite length can look very long to a multiply reflected wave.

(d) In mirror devices, the dominant loss is end loss due to particles traveling along the magnetic field lines. For this process, the confinement time scales only linearly with the length of the device—and with a small coefficient at that (because it does not take long for the high temperature ions in a mirror device to travel along the field lines to the ends). Since the dominant particle loss is through the ends, there is no particular motivation for increasing the radius of mirror devices very much to limit cross-field diffusion. Since the end losses do not decrease rapidly with increasing length, there is not much motivation to increase the length either. We therefore reach the conclusion that if a mirror reactor can be built, a small one will probably be just about as easy to make work as a larger one. Recall that for a Tokamak, on the other hand, the confinement time τ increased as a^2. The interpretation is that, since the particle loss is due to a random walk across the magnetic field, the time required for the particle to escape increases quickly with increasing mirror radius. Thus, there is quite an incentive in Tokamaks to build large machines to increase the confinement time. The conclusion is that in applications where only a small reactor is needed, such as some kind of test reactor, for example, it may make sense to use a mirror device.

(e) Note that the direct converter and injection system improves the energy confinement time without affecting the particle confinement time.

(f) In a plasma, each particle is strongly coupled to its neighbors by Coulomb forces. This interaction or communication permits collective effects, such as waves, in plasmas. For a discussion of when a collection of charged particles can exhibit collective effects and thus be considered to be a plasma, see F. F. Chen, *Introduction to Plasma Physics*, Plenum Press, New York, 1974, pp. 3–12.

(g) Since the electrons and ions stream along together in the expander region and the electrons have a much smaller mass than the ions, most of the energy is carried by the ions with relatively little carried by the electrons. We are not very careful about collecting the electrons with high efficiency, therefore, and use only a single collector electrode for simplicity.

(h) A mirror reactor cannot be "ignited" in the sense of using the energy deposited directly in the plasma by the alphas to heat the plasma and sustain the reaction. Circulation of power is essential. This fact is, of course, discouraging from an economic point of view, although it is too early to tell if it will rule out economic operation of mirror reactors in the long term. One good point about circulating power is that it can be varied to produce a convenient control of the fusion reaction rate. In an ignited reactor, we saw earlier that the operating point is unstable so that control by impurity injection or some clever scheme will be required. Thus, control of an ignited reactor will be a nuisance but it might be much cheaper than recirculating a lot of power.

(i) In an ignited reactor, the input power would be zero so that $Q = \infty$. Also note that in a device with $Q = 1$, the input power is doubled. In general, the output power is $(Q + 1)$ times the input power.

(j) Mirror reactors, with their high T and β, would seem to be ideally suited for reactions such as D-He^3, etc., whose products are primarily charged particles. It seems, however, that the reactor power gain for such reactions is only marginal at best. Stated another way, confinement in mirrors as we now know them does not seem to be quite good enough to consider using reactions other than D-T. But even so, mirrors seem, to the authors at least, to hold more promise for these reactions than Tokamaks as we now know them.

(k) Both the ions and electrons try to escape along the field lines that run out of the reactor region. The electrons move with a much higher thermal velocity, however, and try to run off and leave the ions behind. The resultant charge separation, however, sets up what is called an *ambipolar* potential (recall an earlier note) which works to slow down the electrons and speed up the ions. Multiply charged impurity ions feel a stronger force due to this potential than the singly charged hydrogen isotope ions and are hence preferentially expelled.

(l) These new results were achieved by streaming a low temperature plasma through the ends of the mirror machine. This streaming plasma helps to fill in the holes in velocity space of the main (injected) mirror plasma and hence tends to stabilize the microinstabilities that otherwise result. This plasma also makes a convenient target for the injected neutral beams. The amount of streaming plasma required for stabilization depends on the steepness of the gradients in the plasma. In present experiments (relatively small), a substantial amount of such plasma is necessary—enough to worry us about increasing the already large amount of circulating power in any future mirror reactors. The size of a mirror reactor will be considerably larger than present day experiments, however, although the values of n, T_e and T_i will not be much different. The net result is that the gradients in a mirror reactor will be less than in present experiments. Consequently, only a small quantity of streaming plasma is expected to be needed in a reactor—not enough to affect the energy balance or power flow very much.

For a discussion of the 1975 experimental results, of Q enhancement, and of a proposed new mirror confinement scheme which uses the electrostatic ambipolar potential mentioned earlier, see R. F. Post, "Physics of Mirror Fusion Systems," *Lawrence Livermore Laboratory Report UCID-17312*, November 4, 1976. The paper also appears in the *Proceedings of the Second ANS Meeting on the Technology of Controlled Nuclear Fusion*, Richland, Washington, September 21–23, 1976.

(m) In addition to tweeking various parameters to increase Q, more drastic measures, such as field reversal as in the Astron (see chapter 7) will be investigated.

5 Linear and Toroidal Theta Pinches

In its simplest form, a linear theta-pinch device (see Figure 5–1) consists of a single turn coil which is wrapped around a preformed plasma column ($n \approx 10^{21}$ m^{-3}, $T \approx 1$ eV) in such a way that the plasma more or less fills the volume inside the coil [1]. At some initial time, a capacitor bank charged to a relatively high voltage (typically 10–60 kV) is switched across the coil and the current begins to increase rapidly. Through transformer action, an azimuthal (theta) current is induced in the highly conducting plasma.[a] This current interacts with the axial magnetic field B_z produced by the azimuthal current in the coil to give a radially inward Lorentz force ($q\mathbf{v} \times \mathbf{B}$) on the ions and electrons carrying the current. These particles therefore begin to move toward the axis of the device. Because of the relatively high plasma density and the consequent high collision frequency, the particles in this so-called current sheath sweep the other plasma particles along with them (snowplow model). The net result is that the plasma is quickly squeezed to a smaller radius. In fact, the current sheath travels inward with a typical speed greater than the sound speed in the plasma. Thus, we have induced a kind of radial shock wave in the plasma.[b] It should therefore not be surprising that the process which we've just described causes, in plasma, just the kind of irreversible heating produced by shock waves in gases. As the plasma is compressed more and more, the plasma kinetic pressure increases, the shock wave is partially reflected, and the radius of the plasma begins to increase a little. After some radial bounces, the plasma radius settles down to an equilibrium value corresponding to a plasma density and temperature that give a kinetic pressure which just balances the magnetic pressure due to B_z. At this point, the shock heating or implosion phase of the theta pinch is completed. Consider now the second stage in the heating process. The whole implosion process happens so fast (fraction of a microsecond) that the relatively slowly changing coil current, and hence B_z, have not reached their maximum values (which occur after one fourth of the LC ringing cycle has been completed). As B_z continues to increase toward its maximum, therefore, the increasing magnetic pressure slowly compresses and heats the plasma further. This phase, called adiabatic compression, continues until peak current is reached, usually after a few microseconds.

The plasmas thus generated in a linear theta pinch typically have $n \approx 4 \times 10^{22}$ m^{-3} and $T_i \approx 2$–6 keV. Because of the way in which the plasma is

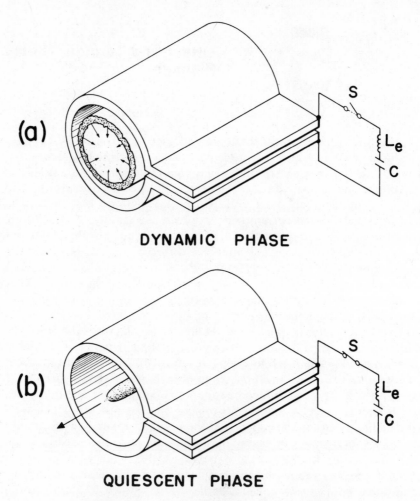

Figure 5–1. A Short Linear Theta Pinch Driven by a Capacitor Bank with Capacitance C and Effective Inductance L_e (courtesy USERDA).

heated and compressed, most of the applied magnetic field is trapped outside the plasma so that this device has high β—in practice $\beta \approx 0.8$. As a consequence, even this hot dense plasma can be confined by a magnetic field of only several Tesla. Perhaps the best news is that no significant MHD instability or enhanced radial diffusion[c] across the applied magnetic field are observed. Shock heating followed by adiabatic compression in a linear theta pinch is therefore clearly a very successful means of obtaining high density, high temperature plasmas.

The obvious difficulty with linear theta pinches is, of course, end losses: most of the particles escape after a time $\tau \sim L/v_i$ where L is the length of the device and $v_i = (kT_i/m_i)^{1/2}$ is the ion thermal speed.[d] If we use $T_i = 6$ keV, $m_i = (m_T + m_D)/2$, $n = 4 \times 10^{22}$ m^{-3}, and, from the Lawson criterion, $n\tau > 10^{20}$ m^{-3}-sec, then we find $L \approx v_i\tau > 1.2$ km. A length of a kilometer or so corresponds to a reactor very much larger than is likely to be desirable.[e] Direct conversion of the energy carried by the particles escaping out the ends (analogous to that proposed for mirror devices) appears to be unattractive because the higher density of the theta-pinch plasma requires a much larger (and hence more expensive) expander region in the converter. Consequently, the favored approach to dealing with the end loss problem has been to bend the theta pinch into a torus. As we discussed earlier, achievement of a toroidal plasma equilibrium requires helical magnetic field lines. If, in considering alternative means [28] of introducing field helicity to achieve an equilibrium we try (1) to keep the geometry as close to linear as possible (long thin plasma: $R/a >> 1$) in order to maintain, hopefully, the favorable stability properties of the linear theta pinch; (2) to use the very successful technique of shock heating followed by adiabatic compression to heat the plasma; and (3) to realize a high β device; then we arrive at the Scyllac configuration depicted in Figure 5–2. This particular configuration can be realized with no toroidal current flowing in the plasma simply by properly shaping the one turn coil.[f]

The Scyllac equilibrium turns out to be weakly MHD unstable to a particular helical distortion of the plasma column.[g] In initial Scyllac experiments, feedback stabilization was used with partial success in stabilizing this mode. It soon became apparent, however, that feedback stabilization would be increasingly difficult and cumbersome in larger machines—especially in reactors. Thus, some sort of alternative stabilization technique was needed. Because the plasma in a theta pinch is pulsed, the use of currents induced in the coil wall (wall stabilization) to oppose displacement of the plasma seems especially appropriate.[h] A property of wall stabilization, however, is that the further the plasma is from the wall, the weaker is the stabilizing effect of the wall. Thus, as the plasma recedes from the wall first during shock heating (or implosion as it is sometimes called) and then during adiabatic compression, wall stabilization becomes a weak effect and cannot stabilize the plasma. To be able to utilize the relatively simple technique of wall stabilization, therefore, we need to find a way to keep the plasma close to the wall even as we heat it by compressing it. Although that sounds like doubletalk, recall that the shock heating (implosion) process is irreversable. As a result, if we decrease the applied B_z after the implosion until the plasma expands to its initial radius, we find that the temperature of the plasma is larger than it was initially.[i] If we now adiabatically compress the plasma by the same factor as we did

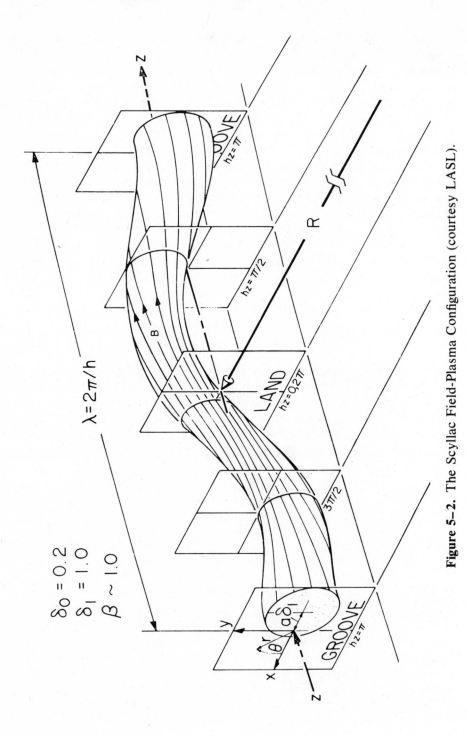

128

Figure 5–2. The Scyllac Field-Plasma Configuration (courtesy LASL).

earlier, with no intermediate expansion, we get the same amount of adiabatic heating as before. The plasma is, however, much closer to the wall since we let it expand to its original size before adiabatically compressing it. Of course, the intermediate expansion and the resultant improved stability are achieved at the cost of reduced net shock heating (since the expanding plasma cools somewhat, although not down to its temperature before shock heating). Thus, there is still a trade-off between heating and stability although it is not as stringent as without the intermediate expansion. A device such as we have just discussed, in which shock heating is followed by expansion of the plasma and then adiabatic compression, is called a staged theta pinch.[j]

The operation of a staged theta pinch is illustrated in Figure 5–3. Notice that there are separate coils for shock heating and adiabatic compression. There are at least three reasons for this arrangement. First, the use of separate coils and capacitor banks permits the two heating processes to be optimized independently. Second, only the energy required for the implosion phase needs to be supplied by an expensive high speed capacitor bank. The adiabatic compression is accomplished with relatively slow but cheap inductive energy storage. Third, only the shock heating coil needs to be wrapped closely around the plasma (so that its inductance will be small enough to permit rapid increases in current) where the neutron flux is high. Fortunately, the shock heating coil can be made thin, to reduce neutron absorption, since the peak shock heating current is relatively small. Note that the compression coil is located outside the blanket. This location means relatively high coil inductance, of course, but adiabatic compression is slow enough that such an inductance is not much of a problem. The field from the compression coil must penetrate the shock heating coil, however. Thus, again, the shock heating coil should be thin. A problem is that if the shock heating coil is too thin, it can wall stabilize only fast instabilities (since the fields from the slower ones will have time to penetrate it). In particular, if the shock heating coil is made just thin enough so that the compression field can penetrate it, then it cannot wall stabilize instabilities that develop on the time scale of (or more slowly than) the compression field. This may be a difficult design constraint. Positioning the compression coil outside the blanket does result in good shielding of the coil. A disadvantage of this location for the compression coil, of course, is that the magnetic field must be maintained not only in the volume occupied by the plasma, but also in that occupied by the blanket.

One module of a conceptual theta-pinch reactor (RTPR) is shown in Figure 5–4. Notice that the compression coil is energized by superconducting storage coils instead of capacitors. This feature results from capacitors simply being too expensive to store the major fraction of the en-

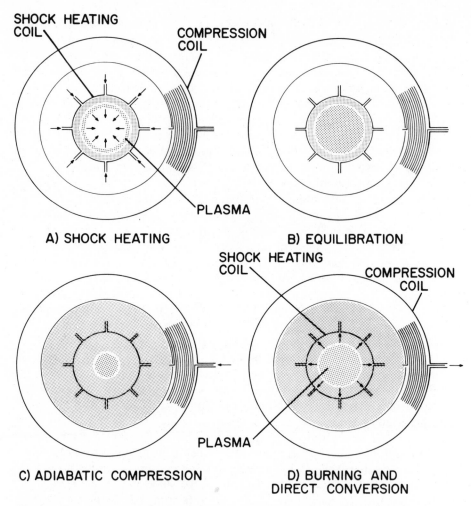

Figure 5–3. Plasma Heating and Burning in a Staged Theta-Pinch Reactor (courtesy LASL).

ergy required by the compression coils in a reactor. Only a relatively small amount of energy is required for shock heating. Consequently, capacitors can be used to energize the shock heating coils.

Recall from chapter 2 that a problem with pulsed reactors is that the energy must be fed into the coils and then be removed and stored cyclicly, with low loss if energy break even is likely to be achieved. Replacing the energy lost in each cycle of operation in a theta-pinch reactor is made simpler, however, by a special kind of direct energy conversion process,

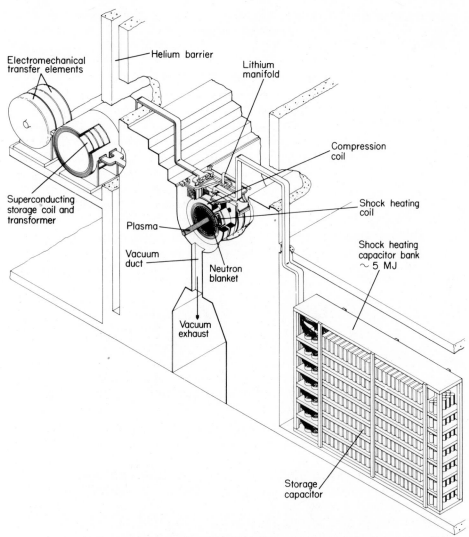

Figure 5–4. Section of the Conceptual RTPR Power Plant Corresponding to One Two-Meter Module (courtesy LASL).

which we now describe. The plasma in the theta pinch is heated to ignition and sustains itself solely by energy deposited in the plasma by alpha particles (no supplementary heating required). Recall also from an earlier discussion (see chapter 2) that the temperature in an ignited reactor increases since the fusion power released increases with temperature more rapidly than the radiation losses. As this happens in a theta-pinch reactor, the plasma pushes the magnetic field lines through the compression coil. The resulting induced voltage drives a current out of the compression coil which, with the use of special circuitry, can be used to recharge the storage inductor.

The parameters of the conceptual RTPR design are listed in Table 5–1.

At the time of writing, a nonstaged version of the Scyllac configuration and a linear version of the staged theta pinch are being investigated experimentally at Los Alamos Scientific Laboratory (LASL).[k] During August 1977, ERDA will review results of plasma confinement (using rf feedback stabilization) and staged heating to determine whether or not they will proceed further with the toroidal theta-pinch program. If an expanded effort is approved for this program, the next experiment will be Staged Scyllac, which is envisioned as a proof-of-principle experiment. If all goes well, a Large Staged Scyllac could come on line in the mid-1980s, a PEPR/ITR in the early 1990s, an EPR in the late 1990s, and a demonstration reactor around 2007. If the toroidal theta-pinch program should be terminated, the program emphasis at LASL would most likely shift either to the toroidal z pinch or the linear theta pinch with the other approach serving as a back up. Several other possibilities, such as the imploding liner (see later discussion), are also being considered.

Notes

(a) The induced current is confined largely to the surface by the skin effect.

(b) This wave is somewhat different from an ordinary shock, however, in that mass is not constant across its front: the mass is swept up by the wave or "magnetic piston" as it is sometimes called. Also note that since the ions and electrons are swept along together at the same speed by the magnetic piston, the ions, being more massive, receive a larger fraction of the energy. Thus, the ions are preferentially heated initially. After an equilibration time of typically a few microseconds, however, the electrons and ions approach the same temperature.

(c) Theoretical investigation of microinstabilities that could cause enhanced radial diffusion in a reactor but which would not occur in present experiments has begun only in recent years. No firm conclusions can yet be drawn from the studies. Earlier it seemed that microinstabilities would not likely be a problem in theta-pinch reactors. The thinking went something like this. The most likely sources of free energy to drive microinstabilities in a theta pinch are the density, temperature, and magnetic field gradients which are an integral part of the confined plasma. Such instabilities, usually termed unstable drift waves or drift instabilities, occur when inhomogeneities cause the particles to drift at speeds which are fast

Table 5–1
Parameters for the RTPR Conceptual Theta-Pinch Reactor

Plasma:
- Beta: 0.8–1.0
- Major Radius R (m): 56
- Plasma radius a: Shock, Burn (m): 0.38, 0.12
- Plasma Aspect Ratio A, Max: 465

Magnetic Fields:
- Shock, Comp. Coil Radii (m): 0.91, 0.94
- Shock Field (T): 1.4
- Compr., Burn Field (T): 11.0
- Helical Poloidal Field (T): 0.8
- Compr. Field Risetime (sec): 0.031
- Supercond. Mag. Energy (GJ): 102
- Shock-Heating Energy (GJ): 0.6

First Wall:
- Radius b (m): 0.5
- Composition: 0.3 mm Ins, 1 mm Nb—1% Zr
- Uncoll. 14-MeV Fluence[a] (n/m²/y): 2.4×10^{25}
- Neutron Curr. Wall Load[a] (MW/m²): 1.9

Neutron blanket:
- Thickness (m): 0.4
- Breeding Ratio: 1.11
- Doubling Time (d): 40
- Material: Li, Be, C, Ins, Nb

Fuel:
- Total T_2 Inventory (kg): 2.2
- T_2 Consumption (kg/d): 0.35
- Total Li Inventory, Tons: 1580
- Li Consumption[a] (kg/d): 37

Burn:
- Shock Ion Density (m⁻³): 2.1×10^{21}
- Burn Initial Ion Density (m⁻³): 2.5×10^{22}
- Shock Temperature (keV): 1.1
- Temperature at Quench (keV): 17
- Burn Time (sec): 0.07
- Burn Parameter $n\tau$ (sec/m³): 8×10^{20}
- Fuel Burnup Fraction: 0.048
- Cooling Time (sec): 1–3

Power Balance:
- Thermal Power P_T (MWt): 3600
- Thermal Conv. eff. (η_τ): 0.56
- Gross El. Power P_E (MWe): 2016
- Circ. Power Fract. (ϵ): 0.13
- Plant Output P_P (MWe): 1750
- Direct Conv. Power P_D (MWe): 350
- Amplification q: 31
- Plant Eff. (η_P): 0.49
- Duty Cycle Time (sec): 10

[a]85% Load Factor

enough to warp the velocity distributions substantially away from Maxwellian. For most of these instabilities, sufficient distortion of the velocity distribution functions occurs only when the particle drift speed is greater than the particle thermal velocity. In a theta-pinch reactor, calculations showed that the drift speeds associated with the inhomogeneities would be smaller than the thermal speed. Thus microinstabilities, and the consequent enhanced particle losses, were not expected to occur. Recent studies have shown, however, that a so-called lower-hybrid-drift instability, for example, can arise even when the drift velocity is less than the thermal velocity [28]. Thus, microinstabilities might turn out to be important in theta-pinch reactors after all, although how important we do not now know.

(d) Accurate, experimentally verified, models for end loss in a linear theta pinch are the subject of current research. Specifically, what value to use for the coefficient k in $\tau = kL/v_i$ is not clear. We use $k = 1$, although this value may be too small and hence to be a pessimistic choice. It may be possible, for example, to use mirrors at the ends to increase k. See, for example, J. P. Freidberg and H. Weitzner, *Nuclear Fusion* 15, 217 (1975). The unfavorable curvature of the magnetic field lines might lead to instabilities if the mirrors are made too strong (mirror ratio too large).

(e) As we mentioned in an earlier note, a fission-fusion hybrid system would ease the energy break-even requirement in a device by using the fusion neutrons to convert U-238 into Pu-239, which is fissionable. The consequent reduction in the $n\tau$ product required for energy break-even could make a linear theta pinch with a length shorter than one kilometer workable as a reactor. The same "trick" can be used, in principle, to reduce the size of most reactor schemes.

(f) The Scyllac configuration can be viewed as a high β version of the Stellarator, a sort of half brother of the Tokamak. Stellarators will be discussed briefly in a later section.

In passing, we note that the Scyllac configuration is not the only high $\beta(\beta > 0.1)$ toroidal configuration possible [28]. A $\beta \lesssim 0.2$ should be achievable in a toroidal device in which the poloidal field required for equilibrium is produced by a toroidal current flowing largely in low density plasma outside the main plasma. This configuration is one form of the so-called "screw pinch." It can be viewed as a high β version of a Tokamak.

Another device that can be viewed as a high β form of the Tokamak is the reversed field, toroidal z pinch in which the poloidal field produced by a fast rising toroidal plasma current compresses, heats, and confines the plasma. Equilibrium and stability are achieved by applying a toroidal field of modest strength, by reversal of the magnetic field at the outside of the plasma, by careful shaping of the current profile, and by wall stabilization (recall discussion of the stabilizing copper shell in Tokamaks). Theoretically, the magnetic field configuration is fully MHD stable for $\beta \lesssim 0.4$. This and other high beta toroidal devices are discussed in chapter 7.

(g) In particular, it is subject to an "$m = 1$ instability," so-called because the distortion varies approximately as cos $(m\theta - 2\pi z/\lambda)$ where θ is the poloidal angle, m is the poloidal wave number, z is a coordinate measured along the minor axis of the torus, and λ is the wavelength of the distortion measured along the minor axis of the torus.

(h) Again, recall the discussion of copper stabilizing shells in Tokamaks.

(i) If the implosion were adiabatic, on the other hand, and the plasma were allowed to expand to its original radius afterwards, then the temperature after the expansion would be just the same as before the implosion.

(j) Even staging may not make wall stabilization work for the $m = 2$ instabilities. The $m = 1$ instability, recall, warps the plasma into a toroidal helix with essentially circular cross section. The development of this instability is hindered by the wall since the magnetic field trapped between the kinks and the wall increases in strength as the kinks move away from the axis toward the wall and hence pushes the plasma back in place. In the $m = 2$ instability, the plasma does not move off axis, but divides into two lobes which form a twisted pair of plasma columns (to see this, sketch $\cos(2\theta - 2\pi z/\lambda)$) along the minor axis of the torus. Since there is no gross movement of the plasma away from the axis, wall stabilization is weak. In experiments up to the present, the minor radius of the plasma columns have been not much larger than the average ion Larmor radius (the radius of the ion orbit in the confining magnetic field). Thus, it is difficult for the two-lobed $m = 2$ instability to form in these experiments since the orbiting particles tend to smear both lobes into one. That is, in present experiments, the $m = 2$ instability is stabilized by finite Larmor radius effects. In a reactor, the plasma minor radius will be much larger than the Larmor radius. Thus, the $m = 2$ instability may be a problem in toroidal theta-pinch reactors since it appears to be only weakly stabilized by wall and finite Larmor radius effects. Note also that staging, which reduces the total adiabatic compression, generally implies lower final plasma density and hence a longer reactor.

(k) The information in this paragraph is taken from *Fusion Power Research and Development Program, Volume III, Five Year Plan, First Draft*, June 4, 1976, prepared by the Division of Magnetic Fusion Energy, USERDA.

6

Laser Pellet Fusion

Recall from our discussion in chapter 2, that the strategy in laser pellet fusion is to use a pulsed laser with enormous peak power to heat a small D-T pellet to fusion temperatures and let it burn while it is confined by its own inertia [1, 29]. Also recall that the energy break even in such a scheme required that a Lawson criterion of the form $n\tau > (n\tau)_{crit}$ be satisfied although $(n\tau)_{crit}$ is larger than the usual 10^{20} m^{-3}-sec because of the relatively low efficiency with which lasers are expected to be able to reinject part of the fusion output energy into the plasma.[a] To consider the constraints this low "circulation efficiency" places on the scheme, consider the simplified energy flow diagram shown in Figure 6–1. In this figure, E_I is the thermal energy input to the pellet from the laser; E_N is the thermonuclear energy produced by the pellet; $Q = E_N/E_I$ is the energy gain of the pellet (analogous to the Q defined as the power gain for mirror machines); E_o is the electric energy out of the thermal cycle where the thermal energy (E_N) is converted to E_o with an efficiency η_T; ϵ is the fraction of power recirculated (the same quantity was defined for mirror devices); η_L is the efficiency with which the laser converts electrical energy to coherent light; E_L is the energy output of the laser; and a is the fraction of the laser energy that appears as thermal energy of the ions and electrons in the plasma. Sustaining a reaction requires the energy loop gain, $Q\eta_T\epsilon\eta_L a$, to be greater than or equal to one:

$$Q \geq \frac{1}{\epsilon a \eta_T \eta_L} \tag{6.1}$$

If we assume optimistically that $a = 1$, $\eta_T = 0.5$, $\eta_L = 0.1$, and for break even that $\epsilon = 1$, then we must have $Q \geq 20$, a value considerably higher than that required in mirror devices. In practice, much of the laser energy may be reflected or may be used up by processes that do not heat the plasma. As a consequence, the fraction of laser energy that ultimately appears as thermal energy of the plasma might be $a \sim 0.1$. If, in addition, we want the fraction of circulating power to be small and choose $\epsilon \sim 0.1$, we see that Qs of over 1000 could be required.

The natural question at this point is, how big can we make Q? The answer to this question depends on how we use the laser energy to heat the pellet, as we will see. For now, however, let us suppose that we use the laser energy to heat a pellet of ordinary solid density uniformly to some

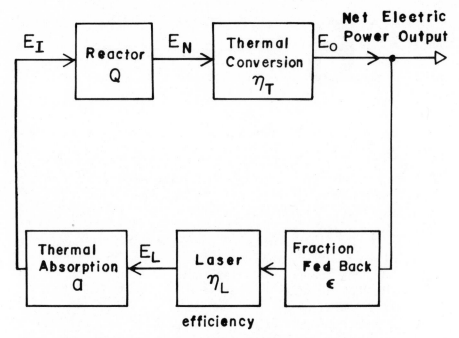

Figure 6–1. Energy Flow Diagram for Laser Pellet Fusion.

temperature T. If m is the mass of the pellet, $m_i \equiv (m_D + m_T)/2$ is the average ion mass, $W_T = 17.58$ MeV is the amount of energy released by each fusion reaction, and f_B is the fraction of the pellet that actually burns up,[b] then the thermonuclear energy released by each pellet is

$$E_N = \frac{1}{2} \frac{m}{m_i} f_B W_T \tag{6.2}$$

where the factor ½ takes into account that two ions are required to produce an energy W_T by fusion. The amount of laser energy that finally ends up as thermal energy in the plasma is

$$E_I = 2 \left[\frac{3}{2} \left(\frac{m}{m_i} \right) kT \right]$$

where the factor of 2 takes into account the thermal energy of the electrons. (We assume $T_e = T_i$ for simplicity.) The energy gain for the uniformly heated pellet is therefore

$$Q \equiv \frac{E_N}{E_I} = \frac{1}{6} \frac{f_B W_T}{kT} \tag{6.3}$$

With $T = 10$ keV, we find

$$Q \approx 300 f_B \qquad (6.4)$$

Since $f_B \leq 1$, by definition, it is clear at this point that the Qs for a homogeneously heated pellet are at best marginal. It is nevertheless important to make f_B as large as possible. From Appendix 6–A, we have for $T = 10$ keV that

$$f_B = \frac{\rho R}{60 + \rho R} \qquad \text{(MKS)} \qquad (6.5)$$

Thus, we can increase the burnup fraction by increasing the radius R of the pellet. For a sufficiently large R, we might be able to achieve break even. As we increase R, however, the volume of the pellet, and hence the size of the laser required, increases rapidly. An alternative approach to increasing f_B, of course, is to increase ρ by compressing the pellet to densities greater than normal solid density. As we indicated in chapter 2, compression of a pellet by laser is indeed possible if the laser energy is used to ablate the surface of the pellet so that the reaction force from the ablated material compresses the pellet by rocket action. In order to look at the effect such compression has on the energy required to heat the pellet to a temperature T, recall that the thermal energy in the pellet is

$$E_I = 2 \left[\frac{3}{2} \left(\frac{m}{m_i} \right) kT \right] \qquad (6.6)$$

and note that we can write

$$m = \frac{4\pi}{3} \frac{(\rho R)^3}{\rho^2} \qquad (6.7)$$

Thus,

$$E_I = \frac{4\pi}{m_i} \left[\frac{(\rho R)^3}{\rho^2} \right] kT \qquad (6.8)$$

In order to realize a given temperature and a given ρR required for break even or for operation of a reactor, note that the thermal energy required decreases as ρ^{-2}. Other things being equal, the laser energy required should decrease by about the same factor. For example, if we seek a temperature of $T = 10$ keV and find that we need $\rho R \approx 30$ kg/m^2 for break even (to realize $f_B \approx 0.33$), then, for the uncompressed pellet, we calculate $E_I = 2.86 \times 10^{12}$J, where we have used the fact that the uncompressed density of solid D-T is 213 kg/m^3. Even after any kind of reasonable extrapolation of today's laser technology, this number for E_I seems hopelessly large. If we were to compress the pellet so that it had a density of 10^6 kg/m^3 (or 10^3 g/cm^3) but decreased its radius so that ρR remains at 30 kg/m^2, then E_I drops to 1.3×10^5 J. Since the laser energy E_L, in practice, will need to be several times as large (perhaps 10) as E_I (because of energy loss to ablating particles, reflections, etc.) even 1.3×10^5 J seems very large, but perhaps not hopelessly so.[c] It is a little discouraging, however, to note that even if we were clever enough to build such a large laser and we then used it to compress and heat a pellet uniformly throughout its volume, that then we would still be stuck with a marginal $Q \equiv E_o/E_I \approx 300 f_B$. Pellet compression, strangely enough, can also be used to solve this problem. Let us see how this could work. Although we did not point it out earlier, compressing the pellet also heats it. If we choose the shape of the laser pulse properly, the center of the pellet can be preferentially heated as spherical shock waves, which are launched by the sudden blast of laser energy at the pellet surface, propagate radially inward and converge. If this central core gets hot enough to ignite, the fusion energy released can drive a supersonic detonation wave which propagates radially outward through the pellet and ignites the pellet material before it has a chance to move. The real advantage of this approach is that the laser is not required to raise the temperature of the entire pellet to ignition. Rather, only enough energy to heat the central core to ignition needs to be supplied. Computer simulations show that in fact the initial hot core comprises only about ten percent of the pellet volume. This result means that the laser needs to supply only about ten percent as much energy to ignite the pellet using this approach as it would need to if the pellet were heated uniformly. This circumstance is, of course, a welcome one as far as reducing the size of the laser required for ignition is concerned. Perhaps the most important point, however, is that the ultimate fusion energy released by the nonuniformly heated pellet is essentially the same as for the uniformly heated pellet (being determined primarily by the pellet mass) so that $Q = E_0/E_I$ has been increased by about ten. This factor of ten, thank goodness, gives us considerably more margin.

There are, of course, difficulties with laser pellet fusion. One of these is to build a laser that is both efficient enough (10%) and big enough. The neodymium glass lasers ($\lambda = 1.06 \times 10^{-6}$ m) that have been widely used in experiments are too inefficient ($< 0.1\%$) for reactor use. Carbon dioxide lasers ($\lambda = 10.6 \times 10^{-6}$ m) are efficient enough ($\sim 10\%$) to be interesting, but have difficulty in supplying the energy in short enough pulses. A new type of laser, affectionately known as the "Brand X Laser," is needed. In addition to high efficiency, the brand X laser preferably should use a gas lasing medium (rather than a solid one like glass) so that permanent optical damage of the medium is not a problem. It also turns out that the brand X laser should have as short a wavelength as possible. To see how this comes about see Appendix 6–B.

Appropriate shaping of the laser pulse to achieve proper compression of the pellet requires peak laser powers considerably higher than the already mind-boggling average laser powers. Optical materials problems may thus be severe. The specific difficulty is that the laser power must be low during the first stages of compression to keep from dumping so much energy into the electrons that they would run toward the center of the pellet, preheat it, and make the succeeding compression more difficult. Keeping the power low during the first stages means much higher power in the later stages, however. This difficulty is alleviated to some degree by using spherical shell pellets, which initially have nothing in the center to be preheated. Plasma instabilities are also a concern. During compression, macroscopic or fluid instabilities triggered by nonuniformities in the pellet or the laser illumination could cause break-up and loss of the pellet plasma. Computer simulations seem to indicate that the compression is completed before these instabilities grow to troublesome amplitudes, however. Microinstabilities are also of concern. One type, called the backscatter instability, could cause enhanced reflection of the incident laser power. Since computer simulations indicate that ideally no more than about ten percent of the incident laser power goes into heating the central core, any enhanced reflection is a cause of concern. How important this instability is in practice remains to be seen.

A worrysome experimentally observed phenomenon which may be caused by microinstabilities is the escape of fast ions from the pellet. Ions with average energies greater than that characteristic of the temperature of the pellet plasma are observed to escape early from the pellet, carrying with them a substantial amount of the incident laser energy. These ions may be accelerated to their high energies by high electric fields produced by microinstabilities at the surface of the pellet where the intense laser fields interact with the plasma. The nature of this phenomenon and whether it can be dealt with in a reactor is not yet understood.

Although high power (0.4×10^{12} W) neodymium glass lasers have already been used to implode D-T gas-filled glass microspheres to about ten times the density of liquid D-T (and produce $\sim 10^7$ thermonuclear neutrons), significant D-T burns are expected to be achieved only upon the completion of SHIVA,[d] the laser part of HELF,[e] which will provide 10 kJ in less than one nanosecond (20×10^{12} W) from twenty simultaneously fired laser chains. SHIVA will have the precise illumination uniformity and high power necessary to compress D-T targets to 10,000 times D-T density and provide neutron yields of as many as 10^{14} per shot. Expected Q values for the HELF system range from 10^{-3} to 10^{-1}. SHIVA design performance goals are listed in Table 6–1. The projected cost for the HELF is 25×10^6. If all goes well, it may be possible to upgrade SHIVA to a power of 200×10^{12} W (200 TW) for a cost as low as $300,000/TW.

The High Energy Gas Laser Facility (HEGLF) at Los Alamos Scientific Laboratory is also scheduled to be completed in 1978. This facility is

Table 6–1
SHIVA Design Performance Goals

Focusable Power	$>20 \times 10^{12}$ W
Focusable Energy	>10 kJ in <1 nsec
Target Illumination Uniformity	$< \pm 10\%$ peak intensity variation
Pulse Shape	30 psec to 2 nsec, simple or complex
Aperture Size	0.63 m²/20 laser chains
Chain Glass Thickness	<2 m
Damage Threshold	>45 kJ at 1 nsec
Linear Aberrations	$< \lambda/2$
Nonlinear Aberrations	$<3 \lambda/2$ at peak power
Beam Alignment	Automatic pointing, focusing, centering
Pulse Synchronization	± 5 psec
Beam Gain Variation	$< \pm 5\%$ peak power

Adapted from "SHIVA: The 10 Kilojoule Nd; glass Laser for the High Energy Laser Facility," University of California, Lawrence Livermore Laboratory, Livermore, California, November 1975.

Table 6–2
HEGLF Power Amplifier Parameters

Length	2.0 m
Pressure	2.5 atm
Gas Mixture	He:N_2:CO_2; 0:1:4
Electron Beam Voltage	-500 kV
Electron Beam Current	50 kA
Main Discharge Voltage	500 kV
Main Discharge Current	500 kA
Discharge Duration	2×10^{-6} sec
Input Electrical Energy	500 kJ
Extractable Optical Energy (Monochromatic)	17 kJ
Pulse Length	0.25–1 nsec
Drive Requirements	10–300 J
Focused Spot Size	400–700 μm diameter

a 6-beam, 10–20-TW, CO_2 laser system with a pulse length of 0.25–1.0 nsec. The total optical aperture, set by the NaCl window damage threshold, is 0.5 m². The parameters for each of the six final power amplifiers are shown in Table 6–2.

Notes

(a) In the discussion of laser pellet fusion, it is customary to talk about the mass density $\rho = m_i n$ (where $m_i = (m_D + m_T)/2$ is the average ion mass) rather than the ion particle density n. In addition, since a pellet of radius R "disassembles" in a time $\tau \sim R/v_s$ (where v_s is the sound speed in the solid pellet), the confinement time is proportional to R. Thus, it is customary to speak of the ρR product rather than the $n\tau$ product. In these terms, an oft-quoted break-even condition is $\rho R > 1$ gram/cm², although the exact value depends on the assumption made about laser efficiency, the details of the absorption of the laser energy, etc.

(b) The burn-up fraction, as f_B is called, increases monotonically as $n\tau$ increases. For $n\tau \approx 10^{20}$ m^{-3}sec, $f_B \approx 0.003$.

(c) For such a pellet, $m = (4\pi/3)(\rho R)^3/\rho^2 = 1.13 \times 10^{-7}$ kg. An uncompressed ($\rho = 213$ kg/m³) D-T pellet with this mass has a radius R of about 0.5 mm. The compressed ($\rho = 10^6$ kg/m³) radius is 0.03 mm. The disassembly time is $\tau = R/4v_s = 6.8 \times 10^{-12}$ sec. The output energy is $E_o = \frac{1}{2}(m/m_i)f_B W_T = 3.8 \times 10^6$ J. The laser power required is $P_L > E_I/\tau = 1.9 \times 10^{16}$ W. Also, $Q = 300\, f_B = 100$.

(d) The information in this paragraph is taken from "SHIVA: The 10 Kilojoule Nd: glass Laser for the High Energy Laser Facility," University of California, Lawrence Livermore Laboratory, Livermore, California, November 1975.

(e) High Energy Laser Facility.

Appendix 6-A

Relation of f_B to $n\tau$ and ρR

Consider a plasma with ion density $n_i = n_D + n_T = n$ before fusion reactions begin [1]. As the reaction proceeds, n_P reaction products (alphas and neutrons) per unit volume are created. Since two reaction products are produced for each two ions consumed in a D-T reaction, then

$$n_i(t) + n_P(t) = n \qquad \text{a constant}$$

For an equal mixture of D and T, the reaction rate R_i is proportional to the product $n_D n_T$.

$$R_i = K n_D n_T = \frac{1}{4} K n_i^2$$

where K, which is a function of temperature[a] alone, has the value $K \approx 10^{-22}$ m³/sec for a temperature $T \approx 10$ keV. Since $\dot{n}_D = \dot{n}_T = \frac{1}{2}\dot{n}_i = -R_i$, then we can write

$$\dot{n}_i = -\frac{1}{2} K n_i^2$$

The burnup fraction $f_B(t)$ is

$$f_B(t) = \frac{n - n_i(t)}{n} = 1 - \frac{n_i(t)}{n}$$

Thus, the differential equation for f_B is

$$\dot{f}_B = \frac{1}{2} K n (1 - f_B)^2$$

If we use as an initial condition that $f_B(\tau = 0) = 0$, this equation can be integrated (assuming T and hence K are constant) to give

$$f_B = \frac{\frac{1}{2} K n \tau}{1 + \frac{1}{2} K n \tau}$$

where τ is the burn time. Note that f_B depends on $n\tau$ and T (through K) and approaches one as $n\tau \to \infty$. If we use the relations $\rho = m_i n$ and[b] $\tau = R/4v_s$, then

---

[a] Actually, $K = \langle \sigma v \rangle$; that is, the product of the D-T fusion reaction cross section and the relative velocity of D and T ions averaged over the velocity distribution function.

[b] We mentioned earlier that $\tau \sim R/v_s$. The coefficient ¼ takes into account spherical geometrical effects.

$$f_B = \frac{\dfrac{K}{8m_i v_s}\rho R}{1 + \dfrac{K}{8m_i v_s}\rho r}$$

At $T = 10\,\text{keV}$, it turns out that $K = 10^{-22}\,\text{m}^3/\text{sec}$ and $v_s = 1.1 \times 10^6\,\text{m/sec}$ so that

$$f_B = \frac{\rho R}{300 + \rho R}$$

In an actual pellet burn, the temperature may increase to 80 keV or so. Detailed computer calculations show that when this temperature rise is taken into account, then the burnup fraction is given approximately by

$$f_B \approx \frac{\rho R}{60 + \rho R}$$

Appendix 6–B

The Plasma Frequency and the Brand X Laser

Consider the propagation of an electromagnetic (EM) wave through a plasma. At low frequencies, the electrons flow in response to the electric field of the wave in much the same way the electrons in a good conductor would. As a result, the plasma exhibits a high conductivity which shorts out the electric field of the wave so that it cannot propagate through the plasma. At sufficiently high frequencies, on the other hand, the electric field of the wave oscillates so rapidly that the electrons, because of their inertia, cannot flow, but merely quiver, as though they were bound, in response to the field. As a result, the plasma looks like a dielectric to a high frequency EM wave. A high frequency wave can therefore propagate through the plasma, although its speed of propagation will be affected by the effective dielectric constant of the plasma. Theory shows that the transition between these two types of plasma behavior is fairly sharp. It occurs at a characteristic frequency ω_p, called the plasma frequency. Thus EM waves with frequency ω such that $\omega < \omega_p$ cannot propagate through the plasma while EM waves with $\omega > \omega_p$ can. It turns out that $\omega_p = (ne^2/m\epsilon_0)^{1/2}$ where n is the electron density in the plasma. A physical interpretation of this relationship is that increasing the electron density provides additional charge carriers, which can keep the electric field shorted out even if the frequency of the wave is increased so that the individual electrons move more sluggishly.

Consider now the application of these results to laser pellet fusion. Far away from the pellet, the plasma density is low and the laser radiation (with frequency ω) propagates freely toward the pellet since $\omega_p < \omega$. As the radiation travels toward the pellet, however, the plasma density increases until finally $\omega_p = \omega$. The electron density at which $\omega_p = \omega$ is called the critical density n_c for the frequency ω. The surface on which $n = n_c$ is called the critical surface. In a spherically symmetric implosion, the critical surface is a sphere centered on the pellet. To a good approximation, the incident laser radiation does not penetrate the critical surface, but is reflected from it. Outside the critical surface, therefore, there is a high density of laser radiation, while inside it there is essentially none. The critical surface therefore represents an interface or boundary between two distinct plasma regions and hence delineates an inhomogeneity in the plasma. We have already noted that inhomogeneities in plasma parameters can drive instabilities. Density gradients can drive drift instabilities, for example. Thus, it is not surprising that the region near the critical sur-

147

face could be the site of plasma instabilities that could disrupt the pellet implosion. The closer the critical surface is to the center of the pellet, however, the higher is the plasma density at the critical surface and hence the higher is the collision frequency there. Thus, the closer the critical surface is to the center of the pellet, the more strongly damped by collisions are any potential instabilities. From our discussion, it is clear that the critical surface can be moved closer to the center of the pellet (where the density is higher) by decreasing the wavelength of the laser. Thus, pellet implosion by a short wavelength laser should be less subject to instabilities than one imploded by a long wavelength laser. As a consequence, the brand X laser should have a short wavelength.

7

Sundry Approaches to Fusion

Although Tokamaks, laser fusion, mirror devices, and theta pinches have received priority funding, other approaches to fusion—some old and some new—may ultimately become more important than these. In this section, we briefly describe several of these alternative approaches.

Bumpy Torus

The bumpy torus concept [30, 31] illustrated in Figure 7–1 is a descendent of the mirror machine in which several magnetic mirrors are placed end to end in a toroidal array to confine particles that would otherwise be lost through the mirrors. As with mirrors, high beta operation should be possible. The poloidal part of the kick that particles (both trapped and untrapped) receive upon encountering a mirror region (recall the discussion of chapter 3), has an effect on the particle orbits somewhat similar to that which a rotational transform has on untrapped particles in a Tokamak or a Stellarator. In particular, encountering the mirror regions moves the particles around the minor circumference of the torus and hence helps to average out the vertical drifts that otherwise would occur because the magnetic field is stronger (in an average sense) on the inside of the torus than on the outside (recall the discussion of chapter 1). Equilibrium between the plasma and the magnetic field can therefore be established. Unfortunately, the unfavorable curvature of the magnetic field lines (convex toward the plasma) in the regions midway between the mirrors (the midplane) means that the equilibrium, as it stands, is not MHD stable (i.e., not minimum-B).[a] Two approaches to achieving a stable equilibrium in the bumpy torus are being investigated. In one, strong radial electric fields applied to the plasma work in conjunction with the basically toroidal magnetic fields in the plasma volume. This drives $\mathbf{E} \times \mathbf{B}$ drifts of the ions and electrons which heat the ions to kinetic temperatures of kilovolts and seem to give stable confinement. An interesting consequence of applying strong external electric fields is that in the usual operating regime ($n_e \approx 10^{17}$ m^{-3}, $T_i \approx$ 1 keV, $B \approx 1$ T), the particle confinement time is independent of the strength B of the applied magnetic field rather than being proportional to B^2 (classical diffusion) or B (Bohm diffusion).

Another approach to achieving stable confinement in the bumpy torus

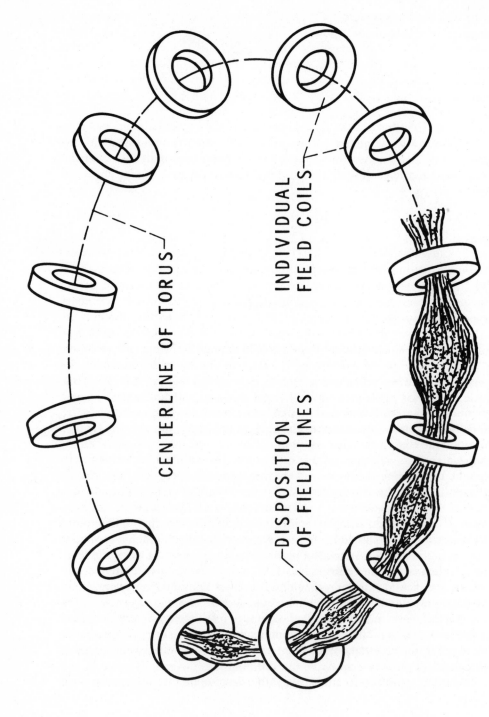

CENTERLINE OF TORUS

INDIVIDUAL FIELD COILS

DISPOSITION OF FIELD LINES

Figure 7–1. The Bumpy Torus. (Courtesy of NASA/Lewis)

is to use high power microwaves to produce high β relativistic electron rings in the midplane between the toroidal field coils. This high beta electron annulus, which is mirror confined, pushes on the magnetic field in the region near the midplane to reverse the unfavorable curvature of the magnetic field lines. The stability of the rings themselves (as opposed to the main body of the plasma) is usually ascribed to some form of line tying of field lines pushed through the (conducting) vacuum vessel by the high beta electron ring.[b]

Potential advantages of the bumpy torus in addition to high beta operation in a closed magnetic field geometry include the possibility of steady state operation and the simple modular nature of the toroidal magnetic field windings.

Astron

A single large relativistic electron ring plays an essential role in the Astron [30, 32] concept, which is illustrated in Figure 7–2. As in the bumpy torus, the relativistic electron ring or e-layer, as it is called in the Astron, is trapped in a magnetic mirror field. In the Astron, however, the current carried by the ring is made sufficiently large that the magnetic field produced by the e-layer is larger than the field applied by the mirror coils. As a result the magnetic field along the axis of the magnetic mirror is reversed in direction compared with that beyond the ring. In fact, the magnetic field lines close around the e-layer and produce a toroidal shaped, absolute minimum-B region in the volume occupied by the electrons in the ring. If D-T fuel is injected into the ring, it is quickly ionized and heated by the relativistic electrons and thus forms a hot stable plasma in the toroidal magnetic well dug out by the e-layer. Thus the e-layer serves both to heat and confine the plasma. Note that although the Astron superficially appears to be a linear machine in Figure 7–2, it is in fact a toroidal device with an elongated cross section, much like a Tokamak with a noncircular cross section. Perhaps the major difference is that the poloidal field in the Astron is produced by the e-layer current rather than by a toroidal current in the plasma. Because the relativistic electrons are very massive, the e-layer turns out to be relatively stiff and consequently relatively stable. In a Tokamak, on the other hand, the poloidal field is generated by the relatively flexible plasma, which becomes unstable, as we saw, if the toroidal current is made too large. Actually, one of the primary purposes of the toroidal magnetic field in a Tokamak is to stiffen the plasma so that it can carry the toroidal current that produces the poloidal field which in turn actually plays a major role in confining the plasma. The size of the toroidal field required to stiffen the Tokamak plasma is so large, in fact,

152

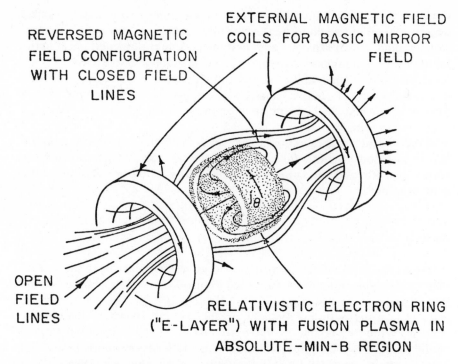

REVERSED MAGNETIC FIELD CONFIGURATION WITH CLOSED FIELD LINES

EXTERNAL MAGNETIC FIELD COILS FOR BASIC MIRROR FIELD

OPEN FIELD LINES

RELATIVISTIC ELECTRON RING ("E-LAYER") WITH FUSION PLASMA IN ABSOLUTE-MIN-B REGION

Figure 7–2. The Astron. (Courtesy of Cornell University)

that the overall β of the Tokamak is, as we saw, very low. No toroidal field is required to stiffen the e-layer.[c] Consequently, β values approaching unity should be possible in the Astron.

The major difficulty with the Astron concept seems to be field reversal. In a series of experiments terminated in 1973, a truly colossal accelerator which could generate ten or more pulses of 6-MeV electrons at a peak current of 600 A with a repetition rate of 960 pulses per second produced an e-layer that gave only fifteen percent reversal of a 0.5 T applied magnetic field [33]. Complete field reversal has been observed for a short time ($\sim 1300\ \mu$ sec) in a subsequent small scale experiment in which the energy of the e-layer was increased by compressing it (by increasing the applied mirror field) after it was injected into the mirror [34]. Compression of the e-layer might therefore be the way to go. It would, of course, preclude steady state operation of the Astron.

Other possible problems with the Astron include synchrotron radiation from the relativistic electrons in the e-layer and the possibility of instability of the e-layer under certain conditions. The synchrotron radiation problem can be dealt with by substituting a p(for proton)-layer for the e-layer [32]. The more massive protons can provide a ring with a given degree of stiffness while moving at a lower velocity than would be necessary for the electrons in an e-layer. The consequent reduction in centripetal acceleration practically eliminates synchrotron radiation.

Toroidal Z Pinch

The toroidal z pinch[d] is a pulsed high density toroidal device in which a fast risetime toroidal current heats, compresses, and confines the plasma with a relatively high beta [35]. In addition to the poloidal magnetic field generated by the toroidal current, a modest toroidal magnetic field generated by external coils is also applied. The toroidal z pinch can therefore also be thought of as a high beta version of the Tokamak. The high beta in these devices is achieved by using a toroidal current large enough to exceed the Kruskal-Shafranov stability limit (see the discussion leading to equation (3.2)). A stable equilibrium is achieved only by reversing the magnetic field at the outside of the plasma, by careful shaping of the current profile, and by wall stabilization. The desired stable profiles are characterized by a hollow pressure profile and a toroidal magnetic field that is reversed in the outer region of the plasma. The resulting stable plasma equilibrium is predicted by MHD theory to exist for betas up to 0.4. Perhaps the major uncertainty, as far as toroidal z pinches are concerned, is whether the programmed profiles can be maintained long enough to achieve adequate confinement before they are severely modified by plasma diffusion and heat conduction.

Stellarator

The Stellarator [36, 37] is a toroidal device in which the rotational transform is produced by currents flowing in external conductors rather than in the plasma, as in Tokamaks. (The toroidal magnetic field is generated by external coils in both Tokamaks and Stellarators.) In particular, the rotational transform is produced by ℓ pairs of conductors which spiral around the torus. These conductors are connected so that the same current flows in each of them, but so that the direction of flow is opposite in adjacent conductors. A sketch of an $\ell = 3$ Stellarator is shown in Figure 7-3. The rotational transform ι due to these conductors (for $\ell > 1$) is proportional to $(\ell - 1)r^{2\ell-4}$, where r is the distance from the minor axis. Note that $\ell = 2$ gives a rotational transform that is constant across the cross section, while $\ell \geq 3$ gives a transform which increases away from the axis. For a given maximum allowed transverse magnetic field, $\ell = 2$ has the advan-

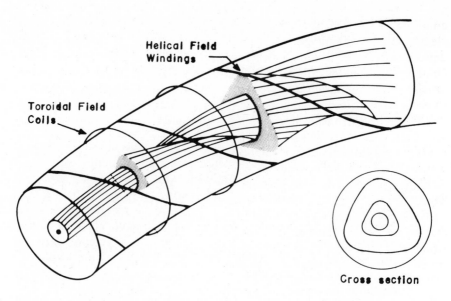

Figure 7–3. The Helical Conductors and Magnetic Surfaces in an $\ell = 3$
Stellarator. Note the shear caused by the radial dependence
of the rotational transform.

tage of providing a large rotational transform throughout the plasma but
has the disadvantage of providing no shear. The $\ell = 3$ case, on the other
hand, provides shear at the expense of giving only small rotational transform near the axis. For $\ell \geq 4$, the volume over which the rotational transform is small increases rapidly. Most Stellarators consequently have $\ell =$
2 or $\ell = 3$.

From Figure 7–3 note that the magnetic surface cross sections in a
Stellarator are not as circular as in an ordinary Tokamak, and that this
noncircular shape rotates as we proceed along the minor axis of the torus.
Thus, the Stellarator, in contrast to the Tokamak, is not axisymmetric
with respect to its minor axis. This lack of symmetry makes confinement
theory for Stellarators more difficult than for Tokamaks [27] and has
sometimes been conjectured to be why the early Stellarators were so infe-
rior to Tokamaks (developed later) which are in principle very similar, except for their axisymmetry.[e] Subsequent experiments indicate, however,
that breaking the axisymmetry in Tokamaks does not substantially affect
their performance. The difference in performance between early Stellarators and early Tokamaks is instead thought to be due to the fact that the
minor radii of plasmas in early Stellarators were relatively smaller than
the minor radii of early Tokamak plasmas. Since we saw in chapter 3,

equation (3.30), that the confinement time $\tau \propto a^2$, the Stellarators were clearly at a disadvantage. More important, however, is that for a given density n, the radial density gradient, dn/dr, is typically larger in a machine with a smaller minor radius than in one with a larger minor radius ($dn/dr \approx n/a$). A larger dn/dr, of course, can drive drift waves unstable. In early Stellarators, therefore, the relatively large density gradients evidently drove drift instabilities which in turn led to enhanced radial diffusion losses. In early Tokamaks, on the other hand, the minor radii were evidently large enough, and hence dn/dr small enough, so that the drift waves were stable. The obvious question at this point is "Why did early Tokamak plasmas have larger minor radii?" The answer follows simply from the fact that the coils for producing the toroidal magnetic fields in early Tokamaks were made roughly the same size as those for the Stellarator experiments that had preceded them. The room inside these coils, which was occupied by the helical windings for producing the rotational transform in Stellarators, was available instead for the plasma in Tokamaks since their rotational transform is produced by currents in the plasma.

Although it has not been experimentally verified at the time of writing, it seems that a big enough Stellarator could perform as well as Tokamaks have. Furthermore, a Stellarator offers the advantages (1) of separate control (optimization) of the heating and confinement functions—the induced toroidal current in Tokamaks plays an important role in both heating and confinement—and (2) of steady state operation—no toroidal current needs to be induced in the plasma by transformer action. Despite these advantages, the increased complexity (and the consequent cost) of Stellarators in comparison to Tokamaks continues to make the Stellarator less popular for fusion experiments than its descendent, the Tokamak. Whether or not Stellarators will be of interest for reactors remains to be seen.

Torsatron

In the Torsatron, a close relative of the Stellarator [30, 38], the rotational transform and shear of the magnetic field lines are achieved with a much simpler conductor geometry. Three separate, equally spaced, conductors, each carrying current in the same direction, spiral around the toroidal volume. These helical windings alone serve the combined function of the toroidal and helical windings in a Stellarator: the poloidal component of the current generates the toroidal magnetic field while the toroidal component of the current generates the poloidal magnetic field. The basic simplicity of the Torsatron magnetic field windings is, of course, very significant from an engineering point of view. Moreover, it is possible, in principle, to design the helical conductors in a Torsatron so that no net mechanical forces act between the individual conductors when they are producing the vacuum magnetic field (no plasma). The supporting structure for such a "force-free" coil configuration needs only to withstand forces that arise from (1) the distortion of the magnetic field away from the force-free vacuum field configuration by finite plasma beta effects and (2) gravity. Since these forces can be much smaller than the unbalanced magnetic forces in a typical non-force-free configuration, the structural design is considerably simplified in the force-free case. Like the Stellarator, the Torsatron is, in principle, a steady state device.

Tormac

In the Tormac [39, 40], as in the Stellarator and the Torsatron, both the toroidal and poloidal confining fields are generated by conductors external to the plasma. The Tormac should therefore be capable of steady state operation. The confinement strategy is quite different in the Tormac, however. As shown in Figure 7–4, toroidal conductors carry currents which produce a transverse magnetic field with favorable curvature toward the plasma (minimum-B configuration). If this field were the only one applied, however, the plasma would be free to escape along the field lines that pass through the cusps in the magnetic field to the walls. In order to "stuff" these cusps, a toroidal field is applied in an attempt to make it necessary for the plasma to diffuse across the toroidal field in order to exit through the cusps. The resultant field is still minimum-B and should be able to confine plasma with a high β. Closed field lines are possible in the main confinement region, however, only if the poloidal magnetic field (whose lines all intersect the vacuum chamber wall) is completely excluded from the plasma. In practice, there is a thin plasma sheath or transition layer in which the poloidal field is not completely excluded. The plasma in the sheath can, by diffusing across the toroidal field and passing through what amounts to a magnetic mirror in the throats of the cusps, gradually escape to the wall (see Figure 7–5). Losing the relatively small amount of plasma in the sheath is not in itself cause for great alarm. What is alarming, however, is that the plasma in the region in which there is no poloidal field is not completely tied to the closed, toroidal field lines. In fact, we saw earlier that a plasma in a purely toroidal field drifts outward across the field lines (along the major radius of the torus) because the magnetic field and hence the magnetic pressure is stronger at the inside of the torus. Thus, as plasma leaves the sheath, it is replaced by plasma drifting in from the main confinement region.[f] The up-shot is that some-

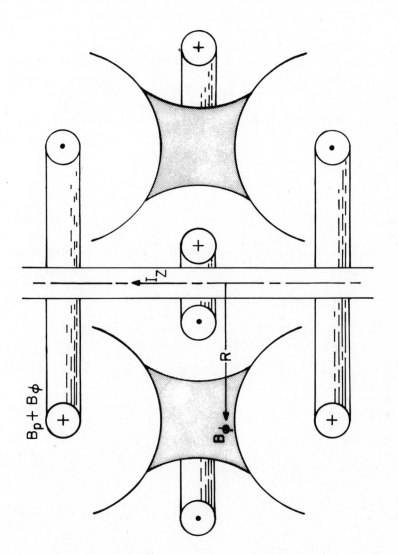

Figure 7–4. Schematic Representation of a Quadrupole Tormac (after Ref. [40]).

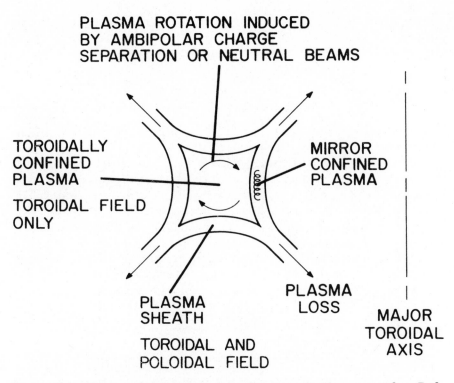

PLASMA ROTATION INDUCED
BY AMBIPOLAR CHARGE
SEPARATION OR NEUTRAL BEAMS

TOROIDALLY
CONFINED
PLASMA

MIRROR
CONFINED
PLASMA

TOROIDAL FIELD
ONLY

PLASMA
SHEATH

PLASMA
LOSS

MAJOR
TOROIDAL
AXIS

TOROIDAL AND
POLOIDAL FIELD

Figure 7–5. Plasma Confinement in a Quadrupole Tormac. (after Ref. [40])

thing must be done to slow cusp losses other than stuffing the cusp with a magnetic field. One approach might be to introduce a rotational transform into the main confinement region. If the transform is produced by a toroidal plasma current, however, β limitations similar to those in a Tokamak are likely to arise. It may be possible to achieve the same effect as a rotational transform (averaging out the particle drifts in a toroidal magnetic field) by simply causing the plasma to rotate about the minor axis of the torus. Such rotation might occur automatically due to a radial (along a minor radius) electric field produced by the space charge that would result if the electrons and ions escaped through the cusps at different rates. Such a field, in conjunction with the toroidal field would produce an $\mathbf{E} \times \mathbf{B}$ drift (see chapter 1) around the minor axis of the torus in the same direction for both ions and electrons. The rotation might be damped, however, since the cusped surface prevents the plasma from rotating as a rigid body. The damping might be overcome by using tangentially injected neutral beams to drive the poloidal rotation.

Another possibility for decreasing the cusp losses is to use microwaves to heat the electrons in the small volume of plasma in the cusps to a temperature of 1 MeV or so [26]. The kinetic pressure of these electrons can retard the escape of plasma through the cusps.

Topolotron

The Topolotron configuration [30, 41], shown in Figure 7–6, arose from highly abstract considerations relating to the topological properties of toroidal magnetic field configurations. The Topolotron exhibits a property known as topological stability, which may also imply improved stability and confinement of a high beta toroidal plasma. The possibility for high β in the Topolotron, as in the Tormac, comes from the fact that the magnetic fields are produced entirely by external conductors. That is, no plasma currents are required for confinement.

The Topolotron has a further interesting property, illustrated in Figure 7–7. The magnetic field lines on the plasma surface are indicated by the arrows in this figure, and they tend to reach limit cycles at the two cusp-like points on the inner circumference of the plasma volume. Whether this limit cycle behavior of the field lines also implies an undesirable piling up of particles at the two cusp points remains to be seen.

Surmac

The strategy of the Surmac (surface magnetic confinement) [40, 42] concept is to generate confining magnetic fields only at the surface of the plasma rather than throughout its entire volume. The volume in which the confining magnetic field must be sustained, and hence the cost of the field, can therefore be reduced considerably. Specifically, the concept involves creating closed magnetic field lines at the surface of the plasma by means of two layers of surface conductors with oppositely directed currents. Figure 7–8 shows a cross section of the plasma together with the magnetic field lines generated by the currents. Note that the magnetic field is essentially zero throughout most of the plasma volume and that the magnetic field lines nearest the main body of the plasma have favorable curvature. The region inside the surface conductors is therefore minimum-B. Only in parts of the relatively small region between the two layers of conductors is the curvature unfavorable. Since the plasma density should be small in the region, the unfavorable curvature may not be too troublesome. Also note that since the confining field does not penetrate the plasma, the β of this scheme is essentially unity.

The basic cross section of Figure 7–8 can be used to generate either a

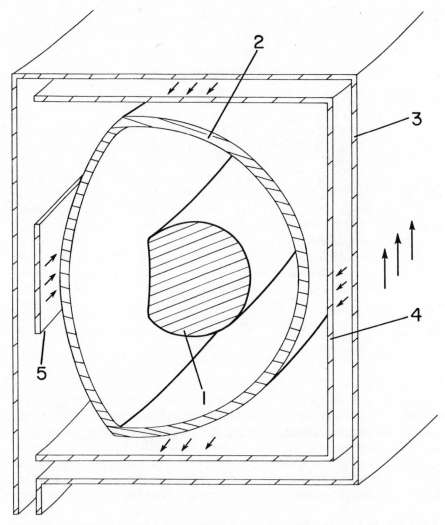

Figure 7–6. The Topolotron Configuration (courtesy Brigham Young University).

toroidal or a cylindrical confinement geometry. In the cylindrical geometry case, the cross section can be slowly tapered to a smaller size at the ends to form "geometrical mirrors" with an effective mirror ratio of approximately 100.

Beyond reducing the magnetic field volume to a small fraction of the volume of the confined plasma perhaps the major beneficial feature of the field free region in a Surmac is the reduction of synchrotron radiation loss

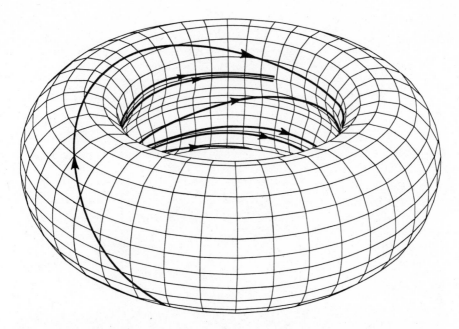

Figure 7–7. Magnetic Field Lines in a Topolotron (courtesy Brigham Young University).

to that of a thin surface layer of plasma. This feature could be especially important for advanced fuel cycles (D-D, D-He^3, etc.), which require higher plasma temperature than D-T fuel. Another possible advantage is that a field free plasma has considerably fewer possible modes of instability than a magnetized plasma. Since the losses occur from the surface of the plasma, however, and the plasma at the surface is magnetized, then whether or not this feature turns out to be an advantage in practice is not yet clear.

It does seem clear that the interior of a Surmac plasma is much more accessible to particle beams than the interior of a magnetized plasma. This feature could permit very effective plasma heating by injecting beams of energetic particles into the Surmac and might permit the plasma potential to be adjusted by the injection of electrons, say. The electric field associated with this potential, working in conjunction with the confining magnetic field, can result in better-than-classical confinement by the magnetic field alone. (Recall the discussion of the bumpy torus.) Specifically, excess electrons injected into the plasma can form an electrostatic potential well for the ions. The electrons are confined by the surface magnetic field.

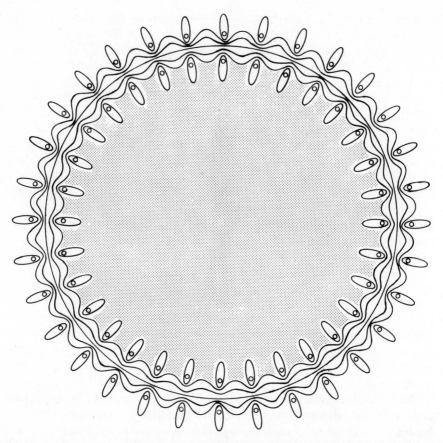

Figure 7–8. A Cross Section of the Plasma Together with the Magnetic Field Lines Generated by the Surface Currents in a Surmac.

Perhaps the major problem apparent with the Surmac, at present, is that the supports for the two layers of surface conductors are in contact with the plasma. This problem could lead to contamination of the plasma as well as deterioration of the supports. The problem is alleviated to some extent by the low density of the plasma in the region where the supports are located. Also, contamination of the plasma could be less of a problem in larger machines where the surface to volume ratio will be smaller.

Electron Beam Pellet Fusion

The electron beam pellet fusion approach [43, 44] is a relative of the laser pellet fusion approach in which energy to compress the fuel pellet by surface ablation is provided by an electron beam rather than by a laser. The electron beam approach offers the advantages of potentially higher efficiencies and of large energy output with presently available technology. As we mentioned in chapter 2, the primary problem in electron beam fusion is in focusing the beam from the fairly large cathode area used to obtain the necessary electron flux to the small size necessary to obtain sufficiently high power density to compress the pellet. One possible technique for focusing the beam is to use a cylindrical cathode from which the emitted electrons converge radially toward a small anode located on the axis of the cylinder. Figure 7–9 shows a relativistic electron beam diode, which represents a further development of this idea. Two ring shaped cathodes emit dual electron beams, which converge toward the pellet at the center of the disc-shaped common anode. The complex self-generated electric and magnetic fields cause the electrons to move over the anode, much as a flat stone skips over the surface of a lake, before finally striking the pellet at the center of the anode and depositing their energy. The pellet and the anode are replaced after every firing. In a reactor, the solid anode would be replaced by a plasma anode. This diode, when driven by the three-conductor transmission disk (radially converging transmission line) shown in Figure 7–10, can present a very low inductance load to a fast, high energy capacitor bank. As a consequence, the energy can be dumped into the pellet very quickly. Specifically, compression of the fuel to about 600 times solid density is expected to occur in less than 20 ns. The transmission disk is enclosed in a water jacket with the water acting as the high energy density dielectric to transfer the electromagnetic pulse from the capacitor bank to the diode. The diode is contained within a spherical chamber with a solid insulator serving to separate the water from the diode vacuum.

A disadvantage of electron beam pellet fusion in comparison with laser pellet fusion is that the fuel pellet will probably be more complex. It must have a sufficiently thick ablator shell to stop the high energy electrons and must have a dense "pusher" material to compress the fuel. Absorption of the incident energy and symmetry and stability of the compression are concerns here just as in laser pellet fusion. Preliminary investigations have not been discouraging, however.

Dense Plasma Focus

The dense plasma focus [45, 46] is another device that makes use of a fast, radially inward flow of particles to achieve a high energy density at the focus of the flow. A dense plasma focus device, shown in Figure 7–11,

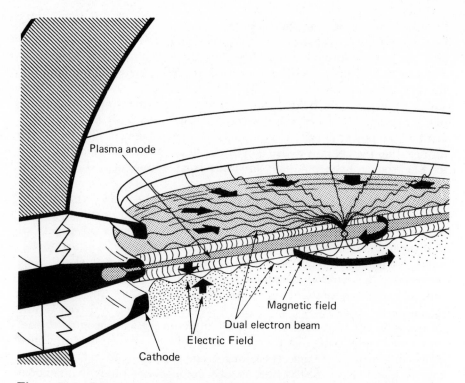

Figure 7–9. Relativistic Electron Beam Diode Especially Designed for Pellet Compression (courtesy Sandia Laboratories).

basically consists of two cylindrical electrodes in a coaxial configuration and a capacitor bank (voltage $\approx$ 20–50 kV; energy $\approx$ 1–1000 kJ, typically 50 kJ). The volume between the electrodes is typically filled with deuterium gas at a pressure of a few torr. When the capacitor bank is switched on, the gas breaks down at the insulator and a radial current starts to flow between the electrodes. This radial current flow produces an azimuthal magnetic field. The radial current flow, in combination with the azimuthal magnetic field generated by it, produces a $\mathbf{J} \times \mathbf{B}$ force on the ions and electrons in the axial direction, away from the insulator. As the current continues to flow, the front of the plasma (ions and electrons) is accelerated along the axis and at the same time ionizes the neutral gas it encounters. When the plasma front reaches the end of the center electrode, the axial momentum of the plasma causes the front to fold over the end and radially compress a small volume of plasma to high density and temperature. This volume of plasma is the dense plasma focus. Plasma densities in the focus of up to 10^{26} m^{-3} have been reported. The focus forms within a

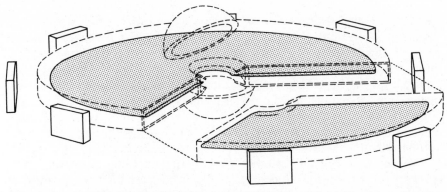

Figure 7–10. The Three Conductor Transmission Disk. The boxes represent the high energy fast capacitor bank (courtesy Sandia Laboratories).

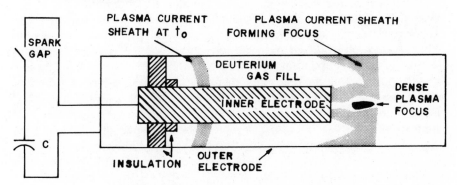

Figure 7–11. A Dense Plasma Focus.

few microseconds. Violent MHD or macroscopic fluid instabilities develop after several nanoseconds. Neutrons are observed for longer times (~100 nsec), however.

Depending upon the details of the device geometry, the particular values of parameters subject to external control (electrode size and geometry, filling pressure, capacitor bank voltage and energy, etc.), the price of rice in China and other factors, the plasma focus can produce either thermal plasmas characterized by ion temperatures up to 9 keV or deuteron beams (presumably accelerated to high energies by the electric fields associated with fluid instabilities) with energies up to 1 MeV. Soft x-radiation is sufficiently intense to produce images of small biological objects with a single shot. Bank energy is converted into hard x-rays in the 100-kV range with an efficiency ~5 × 10⁻⁴. X-ray quanta with energies up to 1

MeV have been observed. Development of theoretical understanding of the detailed physical processes in a plasma focus has been severely hampered by the difficulty of making detailed measurements on such wee (~1 mm diameter), fleeting, plasmas. As a consequence, the available scaling laws are empirical in nature. For example, it seems that the number of neutrons produced per shot increases roughly as E^2 where E is the energy stored in the capacitor bank. This result, of course, indicates very favorable scaling, but cannot be used for extrapolation to very large machines without a better theoretical understanding of its origins. It is also important that experiments indicate that many if not most of the neutrons result from fusion reactions produced by collisions between high energy ions in a beam and lower energy ions in the main background plasma. As we argued in chapter 1, such a fusion mechanism is unlikely to be interesting for energy production. It appears, therefore, that application of the dense plasma focus for fusion purposes is limited to a possible source of neutrons for testing reactor materials, and the like. For example a 250-kJ device can produce more than 10^{12} neutrons per shot. Other possible applications of a dense plasma focus as a neutron source include pulsed neutron photography, pulsed neutron activation analysis, nuclear weapons simulation, and neutron therapy. Applications proposed for the light and x-rays emitted from the focus include generating highly stripped atoms for spectroscopic purposes, serving as a pump source for lasers, providing x-ray diagnostics in medicine, and producing x-rays for nuclear weapons simulation.

Imploding Liners

Imploding liners [40], like laser and electron beam pellet fusion devices, try to exploit the $n\tau$ trade-off possible in satisfying the Lawson criterion by achieving large n so that only a small τ is required. In comparison with the laser and electron beam approaches, imploding liner devices operate with lower $n(\sim 10^{26}\,\text{m}^{-3})$ and hence require longer confinement, but offer the advantage of not requiring sophisticated lasers or electron beam sources.

The basic imploding liner approach is to generate a low density plasma inside a cylindrical conducting (usually liquid metal) liner and then to apply radially inward mechanical or electromagnetic forces to compress or implode the liner. In the early stages of the liner implosion, the liner gains considerable energy and momentum from the applied radially inward forces without feeling much radially outward force from the plasma inside it. During the final stages of implosion, the plasma exerts considerable outward radial pressure on the liner. By this time, however, the liner has gained substantial energy and momentum from the applied forces and can cause very large compression of the plasma before being repelled. The liner, therefore, serves as a kind of storage tank for energy and momentum, which can be filled more or less gradually and then emptied quickly when the need arises.

Magnetizing the plasma before imploding the liner has both advantages and disadvantages. The magnetic field can serve as a partial barrier between the liner and the plasma to inhibit cross-field particle diffusion and hence provide thermal insulation between the hot plasma and the cooler liner. This helps prevent particles from the liner from penetrating the plasma as impurities. Since magnetic fields of up to a few hundred tesla can be achieved at peak compression, considerable isolation between the liner and the plasma is possible. As this enormous magnetic field begins to diffuse into the conducting liner during implosion, however, it induces surface currents, which can vaporize the liner surface much more quickly than the direct plasma bombardment of the liner that would occur if there were no magnetic field present. Thus, to what extent, if any, a magnetic field reduces the net influx of impurities into the plasma from the liner is not yet clear. Even if the magnetic field did not reduce the influx of impurities significantly, however, the reduction of heat conduction to the wall alone might make it worthwhile to use a magnetized plasma.

A definite advantage of using an unmagnetized plasma inside the liner, on the other hand, is improved plasma stability. For in this case, the plasma is supported directly by the relatively massive liner, which exhibits considerable inertia to the motions associated with instabilities in contrast to the massless magnetic field that confines the magnetized plasma. As a result, the growth rates of instabilities are considerably reduced. Other instabilities simply cannot occur without a magnetic field (recall the discussion of Surmac).

If a magnetized plasma is utilized, an appropriate magnetic field configuration must be chosen. A particularly simple configuration is a constant, uniform field directed along the axis of the liner. Such a plasma could be prepared by theta-pinch action. In this configuration, heat conduction out of the ends (along the magnetic field lines) is unabated, however. If, on the other hand, an azimuthal magnetic field is generated by passing a current through the plasma parallel to the liner axis (z pinch), then thermal conduction out of the ends, as well as to the liner, can occur only by cross-field diffusion, a relatively slow process, as we have seen. It may be, therefore, that the azimuthal field configuration (z pinch) is preferable to the axial field configuration (θ pinch).

The forces applied to implode the liner can be mechanical or electromagnetic in nature. In the later case, the liner and plasma are surrounded by a coil whose axis coincides with that of the liner. When a strong current is passed through the coil, a strong magnetic field is generated in the region between the coil and the liner. The consequent magnetic pressure causes the liner to implode. In this arrangement, the liner is imploded in much the same way that the plasma is imploded in a theta pinch. Although the energy required to implode the liner could be provided by a capacitor bank, cost considerations make it likely that inductive energy storage would be used in large imploding liner devices.[g] Figure 7–12 shows a simple inductive energy storage circuit with an inductive load. Initially the switch S is closed, and the storage inductor L_0 has been charged so that a current I_0 is flowing through it. The energy stored in L_0 is $\frac{1}{2} L_0 I_0^2$. When S is opened, the current I_0 is transferred from the switch to the inductive load L. Unless $L \ll L_0$, a very large voltage (di/dt and hence $V = L \, di/dt$ would be very large across L) appears across S during switching. This voltage makes it difficult for S to open properly. Having a liner device as a load (instead of an ordinary inductor) alleviates this problem to some degree. In the early stage of the implosion, the liner (which can be viewed as the secondary of a transformer in which the coil is the primary) and the coil are closely coupled so that the apparent inductance of the coil is low. The initial voltage across S can therefore be relatively small. As the liner implodes, however, the coupling between the liner and the coil becomes weaker so that the apparent inductance increases. This larger inductance improves the efficiency with which energy is transferred from the storage inductor to the liner device.

Ideally, an electromagnetically driven liner system functions as (1) a mechanical energy storage system in which electrical driving energy is stored as kinetic energy of the imploding liner; (2) a transducer which transforms this kinetic energy into plasma thermal energy by compression; (3) an inertial containment device; (4) a transducer which transforms the thermal energy of the plasma (hopefully increased by the release of fusion energy) into kinetic energy of the liner as it is accelerated outward by the plasma pressure; and (5) an energy recovery device which transforms its kinetic energy back to electricity by pushing against the magnetic field that was applied to initiate the implosion.

Out of several practical difficulties with electromagnetic implosion, we might single out ohmic dissipation in the liner as being one that can affect the energy balance unfavorably in a very direct way. Another possible difficulty with implosion of the liner is liner stability. Even if the liner is initially solid, it quickly melts (because of internal friction, if nothing else). The basic problem with liquid liners is the Rayleigh-Taylor instability that prevents a heavier fluid from stably resting on a lighter fluid. Experiments show, for example, that water in an inverted glass tumbler is not stably supported by the atmosphere even though the force exerted on the surface of the water by atmospheric pressure is greater than the

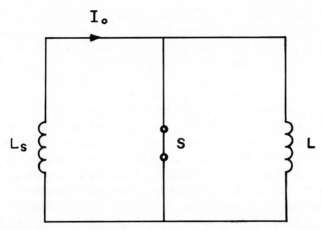

Figure 7–12. A Simple Inductive Energy Storage System.

weight of the water. That is, when you turn a glass of water upside down, the water comes out even if you didn't expect it to. The Rayleigh-Taylor instability between a light and a heavy fluid can occur not only when the fluids are subject to forces due to gravity, but also when the force is due to acceleration or deceleration. That is, the origin of the force acting on the two fluids is not important as far as the occurrence of the instability is concerned. The important thing to remember is that the unstable situation occurs when, with respect to the direction of the force, the light fluid is in front of the heavy fluid. If the force arises from acceleration, then it is in the direction opposite to the acceleration (action-reaction). For example, when an elevator accelerates upward, you feel a downward force. In terms of acceleration of the fluids, therefore, we can say that the unstable condition occurs when, with respect to the direction of the acceleration, the heavy fluid is in front of the light fluid. Now let's apply this result to the liner implosion process.

In the early stages of implosion, the liquid liner and the plasma are accelerating inward. Since the plasma (the lighter fluid) is in front of the liner (the heavier fluid), the liner-plasma interface is stable during this stage. As the implosion progresses, however, the plasma pressure increases and begins to decelerate the inward motion of the liner and the plasma. The acceleration at this point reverses and points outward. The implosion process then becomes unstable, and confinement of the plasma is lost. What we need is some kind of trick to keep the acceleration inward at all times during the implosion. A simple way to do this is to have the cylindrical liquid liner rotate about its axis during the implosion. The rotation gives a centripetal (inward) acceleration, which increases as the liner is compressed. It turns out that relatively modest initial rotation of the liner gives enough centripetal acceleration to make the net acceleration inward, and hence stable, throughout the implosion. The initial liner rotation is also helpful in forming a symmetrical hollow liquid liner to start with.

An alternative to electromagnetic implosion of the liner is mechanical implosion. In this case, the liquid liner is held against the inside surface of a rigid rotating cylinder block by centrifugal force. Pistons in the rigid cylinder are driven inward (by steam or gas pressure, electromechanical forces, etc.) to implode the liner. High implosion speeds can be achieved even with relatively slow mechanical drivers for the pistons because the velocity of the inside surface of the liner is multiplied during compression as the essentially incompressible liner thickens. (The same effect occurs during electromagnetic implosion.) This thickening during implosion may be handy in providing a neutron blanket and a radiation shield for the rotating cylinder block, pistons, and the like. The rotation of the liner in the so-called captive liner system we've just discussed serves the same stabilizing function as in the electromagnetic implosion or free liner system we discussed earlier. In both cases, it may be possible to use the energy initially stored in the liner rotation to drive the liner implosion and hence reduce the requirement for costly external (inductive or capacitive) energy storage. Such systems are called self-driven liner systems [48].

Although it is easy to point out potential difficulties (recycling of liners, ohmic dissipation, end losses, conversion, etc.), more experimental work is required for a useful evaluation of the concept.

Laser Heated Solenoid Fusion

Laser heated solenoid fusion devices [49, 50] can be viewed as variants of the long linear staged θ pinch in which the shock heating stage is replaced by laser heating of the plasma using a beam that propagates along the axis of the machine. This substitution results in a device which is considerably different from a linear theta pinch, however. The only kind of laser presently available that seems scalable to adequate power and has reasonable efficiency (up to 30%) is the carbon dioxide laser. (The CO laser might eventually be competitive, however.) The characteristic length for absorption of the CO_2 laser beam in a plasma by inverse bremsstrahlung[h] is

$$\ell \approx 3.24 \times 10^{48} \, T_e^{3/2}/n_e^2 \text{ meters}$$

where T_e is the electron temperature in keV and n_e is the electron density (m^{-3}).[i] For $T_e \approx 10$ keV, $\ell \sim 10^{50}/n_e^2$. Clearly this absorption length should be made shorter than the length of the plasma in order to extract most of the energy from the beam as it passes through the plasma. We have already argued in the case of the linear theta pinch that the length of such a linear device must be about 1 km or so to limit end losses to an acceptable value without some kind of end-stopper (mirrors, etc.). In order to make the characteristic absorption length $\ell \sim 10^{50}/n_e^2 < 1000$ m, we must choose $n_e \sim 10^{24}$ m^{-3}. Such a large value for the density means, of course, a large value for the magnetic field. To keep the magnetic field as small as possible, suppose that we operate with $\beta = 1$ so that $B_{min} = (4\mu_0 n_e kT)^{1/2}$. If we choose $T = 10$ keV, we find $B_{min} \approx 61$ T, a very large magnetic field.[j] If we want to reduce the magnetic field as much as possible, we might choose $T = 5$ keV in which case $B_{min} \approx 43$ T. Even this value corresponds to a magnetic field of prodigious strength. The only thing standing between us and abject despair at this point is the hope that the relatively high ion density ($\sim 10^{24}$ m^{-3}) will let us achieve a reasonable power per unit length with only a small diameter (~ 1 cm) plasma. If this were the case, the magnetic field could be produced by coils with a small bore. Such coils, of course, have a much better chance of withstanding the enormous forces required to produce 45 T or so than coils with larger bores. To pursue this possibility further, we note that to meet the Lawson criterion in a plasma with $n_e \sim 10^{24}$ m^{-3} we must require τ to be something like a millisecond for break even.[k] In a reactor, τ should probably be about 10 ms. It turns out that the fusion power density for $n_D = n_T = n_e/2 \sim 5 \times 10^{23}$ m^{-3} and $T = 5$ keV is roughly 10^{12} W/m^3.

If we use a plasma with a cross sectional area of 1 cm^2 (10^{-4} m^2) and a 1-km length and maintain it at $T = 5$ keV for 10 msec, then fusion energy in the amount of 10^9 J is released during the pulse. We want to choose the radius of the first wall around the plasma to be as small as possible in order to permit the bore of the magnets to be small but large enough to allow compression of the plasma to heat it.[l] If we choose the diameter of the wall to be about 3 cm, then its circumference is about 10 cm so that its total wall area is about 100 m^2. If t_0 is the time between pulses, then the average wall loading (W/m^2) is 10^9J/(100 m^2)t_0. To limit the average wall loading to 10^6 W/m^2, we take the time between pulses to be about 10 sec. In this case, the average fusion power output is about 100 MW. Thus, it does seem possible to use a small bore magnet and still produce significant if not plentiful fusion power output. We'll discuss increasing the output power a little later.

We now turn to the problem of providing the magnetic field of about 45 T required for operation with $T = 5$ keV. We would like to provide as much of the magnetic field as possible by means of a superconducting solenoid although we must pulse on some of the field to compress the plasma. At present, it appears that 20 T is the maximum field available from such solenoids. Thus the pulsed coil must supply about 25 T.

Compression heating of the plasma is important for two reasons. First, it is more efficient and less costly than laser heating. Second, any heating that comes from compression reduces the necessary size of the laser—a very desirable result. Laser heating is essential (as was shock heating in the staged theta pinch) nevertheless since (1) drastic compression of the plasma makes wall stabilization of plasma instabilities ineffective and (2) compression only multiplies (as we will see later) the energy already present in the plasma so that preheating of the plasma before compression is essential. In addition to heating the plasma by adiabatic compression, the pulsed coil, in fact, would also serve the important purpose of compressing the plasma away from the walls to cut down on the plasma heat loss due to radial thermal conduction to the walls.

A schematic of a possible hybrid arrangement of pulsed and superconducting coils is shown in Figure 7–13. In this configuration, several (perhaps ten) plasma tube-pulsed coil assemblies are placed in a sort of Gatling gun arrangement inside the superconducting coil. This approach would permit each of the assemblies to be fired once every ten seconds, say, while the reactor as a whole would fire several times (ten times if there are ten assemblies) as fast. As a result, the power out of the reactor will be multiplied by the same factor. If ten assemblies of the type we considered earlier were used, the fusion power output of the reactor would become 1000 MW instead of 100 MW. The electrical power output would be something like 400 MW, assuming an electrical conversion efficiency of about forty percent.

A question that naturally arises at this point is "How big does the laser need to be?" Well, the total energy required to heat the plasma to a temperature T is $E = 2Vn_e((3/2)kT)$ where V is the volume and we have taken $T_e = T_i = T$. With $V = 0.1$ m^3, $T = 5$ keV and $n_e = 10^{24}$ m^{-3}, we find $E \approx 2.2 \times 10^8$ J. But how much of this can be supplied by adiabatic compression and how much must be supplied by the laser? If the compression is adiabatic, then $p_0 V_0^\gamma = pV^\gamma$ where p_0 and V_0 are, respectively, the pressure and volume before compression and $\gamma = 5/3$ for slow processes (like adiabatic compression) in a plasma. Using the fact that we can write $E = (3/2)pV$ since $p = 2n_e kT$ and the fact that $V \propto r^2$ where r is the plasma radius, then we find

$$E/E_0 = (V_0/V)^{\gamma-1} = (r_0/r)^{2(\gamma-1)} = (r_0/r)^{4/3}$$

where E_0 and r_0 are respectively the plasma energy and radius before compression. If the initial diameter is 3 cm (the wall diameter) and the final diameter is 1.12 cm (to give $\pi r^2 \approx 1$ cm^2), then the energy gain, E/E_0, due to compression $(3/1.12)^{4/3} = 3.7$. Thus, if E_0 is the plasma energy before

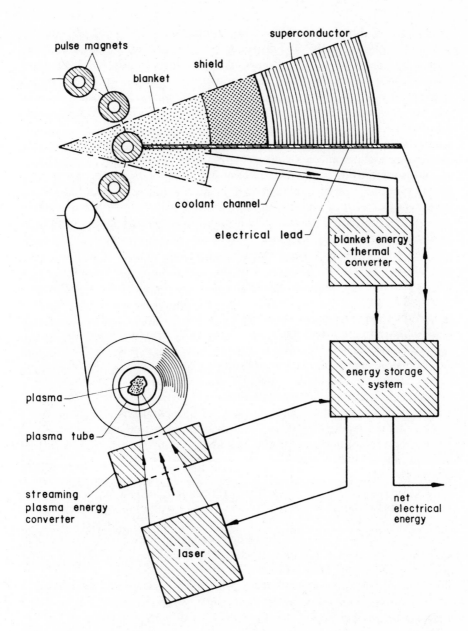

Figure 7–13. Schematic Diagram of a Laser Heated Solenoid Power Reactor (courtesy Electric Power Research Institute).

compression, the plasma energy after compression is $E = 3.7\,E_0$. But we must have $E = 2.2 \times 10^8$ J. This requires $E_0 \approx 60 \times 10^6$ J. But the energy before compression must be supplied by the laser. Thus, the laser must supply about 60 MJ, assuming reflections, instabilities, and the like do not prevent any of the laser energy from being absorbed. Although this is a huge laser, it does seem possible to scale up present CO_2 laser technology to this size.

Note that we assumed $\tau = 10$ ms although for a linear device one kilometer long with no mirrors, and so forth, the end loss time is, as we saw in the discussion of theta pinches, only about 1 ms. Thus, realizing a reactor of the type we've just described will require some sort of end stoppers. These may consist of multiple mirrors at the ends of the device or even of solid plugs (perhaps D-T ice) which build up insulating layers of neutral gas and cold plasma by ablation. The success of end stoppering is crucial, but has not yet been demonstrated.[m]

Another potential difficulty with this approach is the possibility that the laser beam may wander out of the very skinny plasma column over a 1-km path, although it seems that self-regulating mechanisms are present which tend to keep the beam centered in the plasma.

Compared with laser pellet fusion, the laser heated solenoid approach has the advantages of already having the laser that is needed (no Brand X laser must be invented) and of not requiring the careful laser pulse shaping and focusing needed in laser pellet fusion. Compared to the theta-pinch concept, the small bore of the laser heated solenoid together with the fact that part of the magnetic field is supplied by superconducting solenoids means that less energy needs to be applied to the pulsed coils for adiabatic compression. Compared with the Tokamak, the laser heated solenoid offers the potential of a much simpler reactor design because of its simple geometry.

At present, USERDA is sponsoring a plasma physics proof of concept experiment for the laser heated solenoid at Mathematical Sciences Northwest, Inc., at Bellevue, Washington. Quoting now from Reference [49], "In this experiment, the following demonstrations are projected: (1) efficient laser beam trapping in a slender plasma column with a length to diameter ratio greater than 1000, (2) plasma heating to equilibrated electron and ion temperatures in excess of 1 keV, and (3) design and development of the subsystems relevant to such an experiment, including a 10-kJ [CO_2], pulsed laser and a 3-meter, 300-kG solenoidal magnet."

Notes

(a) At this point, one might wonder whether this problem might be avoided by cascading the minimum-*B* mirrors of Figures 4–1 or 4–2 (instead of the simple mirrors shown in Figure 7–1) with a hope of achieving simultaneously a closed and minimum-*B* geometry. That a configuration which is simultaneously closed and minimum-*B* and has no currents flowing in conductors within the plasma does not exist, however, has been shown by J. D. Jukes, *J. Nucl. Energy* C **6**, 84 (1964). Even so, the performance of the toroidal array of minimum-*B* mirrors might be better than that of a corresponding array of simple mirrors.

(b) Magnetic field lines embedded in a perfect conductor are "frozen" in or "tied" to the conductor. In a good conductor, they can move only slowly. Because a plasma is a very good conductor (typically much better than copper), the field lines and the plasma tend to be locked together. If the plasma is tied to field lines which are in turn tied to a sturdy wall, it is difficult for the plasma to move. Consequently, MHD or macroscopic instabilities tend to be suppressed by line tying.

(c) An even stiffer means to carry the toroidal current is a metal ring immersed in a toroidal plasma. Machines which use this approach, called internal ring devices, are probably unsuitable for reactor purposes because neutrons, gamma ray backshine, and plasma bombardment would damage the ring and inject high-*Z* elements into the plasma. Such devices have been very important in understanding the physics of toroidal plasmas, however. The e-layer, although not as stiff as a metal ring, has the advantage of being unaffected by radiation, etc.

(d) In this context, the *z* axis is the minor axis of the torus.

(e) Except for modifications due to nonaxisymmetry, the theory for Stellarators is similar to that for Tokamaks; there are trapped and untrapped particles, bananas, etc. The β limit for Stellarators is $\beta \ll (8/\pi^2)(a/R)$, not much different than for a Tokamak, although it arises from somewhat different considerations (see Ref. [6], pp. 302–304).

(f) If the cusp loss rate can be controlled, the cusps might be useful as divertors.

(g) It might also be possible to use a homopolar generator [47]. In its simplest form, a homopolar generator consists of a conducting flywheel immersed in a constant uniform magnetic field directed along the axis of the flywheel. This magnetic field in conjunction with the azimuthal rotation of the flywheel produces a radial electric field. Thus a voltage appears between electrodes connected (by brushes) to the shaft and the edge of the flywheel, respectively. Homopolar generators are a relatively fast (low inductance) means of converting the kinetic energy of flywheels to electricity. In terms of Joules/$, flywheels are much cheaper than inductors and capacitors (and batteries). The energy is usually not convertible at fast rates, however.

(h) Recall that bremsstrahlung occurs when an electron is accelerated by an ion during a "collision" and hence radiates. From a quantum point of view, bremsstrahlung occurs when an electron collides with an ion and loses energy by emitting a photon. The corresponding view of inverse bremsstrahlung is that an electron collides with an ion, absorbs a photon, and hence gains energy. To view inverse bremsstrahlung from a classical point of view, we note that an electron in the presence of a plane electromagnetic wave (as from a laser), applied at some initial time, begins to oscillate in response to the oscillating electric field. (The response to the magnetic field of the wave is small except for electrons with very high velocity.) As time passes, the amplitude of the electron oscillations and hence the oscillatory kinetic energy of the electrons approaches a steady state value. If at some point in the oscillatory cycle, the electron collides with an ion, then in the process of bouncing off, part of the oscillatory kinetic energy can be converted into translational energy (equivalent to thermal energy). As long as the translational velocity is small compared to the speed of light, the basic interaction process between the electron and the electromagnetic wave is unaffected by the translational velocity. Thus, after the collision, the electron again builds up oscillatory energy by absorbing energy from the wave. Repetition of this process clearly provides a means of transforming the higher ordered energy of the electromagnetic wave into thermal motion of the particles.

(i) A long wavelength laser such as the CO_2 laser has the advantage in laser heated solenoid fusion that the absorption length ℓ is smaller than for shorter wavelength lasers, such as CO ($\lambda = 5 \times 10^{-6}$ m). Specifically, $\ell \propto \lambda^{-2}$.

(j) Anomalously large absorption (perhaps due to instabilities driven by the intense laser radiation) or a configuration in which the laser beam bounces back and forth through the plasma several times (multipassing) could lower the required value of n_e and hence permit B to be smaller.

(k) Recall from earlier discussions that if the Lawson criterion is written as $n\tau > (n\tau)_{crit}$, then $(n\tau)_{crit}$ for a laser heated plasma must be somewhat greater than the 10^{20} m^{-3}-sec (which we usually use) because of the relatively low efficiency for even CO_2 lasers. Using $T = 5$ keV instead of 10 keV increases $(n\tau)_{crit}$ even more. This fact is why we need a 1-km device to break even though the ion density is 10 times higher than in a linear θ pinch.

(l) The absolute minimum value of the wall radius r_W, is $r_W = r_p + 2r_\alpha$ where r_p is the radius of the plasma and r_α is the Larmor radius (orbit radius in a magnetic field) for the alpha particles. If the wall radius is smaller than this, the wall can be severely damaged by the high energy alpha particles. For $B = 45$ T, $r_\alpha \approx 6$ mm.

(m) The requirements for end stoppering and practically everything else can be considerably reduced by using fission-fusion hybrid schemes.

References

1. F. L. Ribe, "Fusion Reactor Systems," *Reviews of Modern Physics,* 47, 7 (1975).

2. R. F. Post and F. L. Ribe, "Fusion Reactors as Future Energy Sources," *Science,* 186, 397 (1974).

3. Don Steiner, "The Technological Requirements for Power by Fusion," *Proc. IEEE,* 63, 1568 (1975).

4. Robert G. Mills, "The Promise of Controlled Fusion," *IEEE Spectrum,* 8, 24 (1971).

5. David J. Rose, "Controlled Nuclear Fusion: Status and Outlook," *Science,* 172, 797 (1971).

6. Samuel Glasstone and Ralph H. Lovberg, *Controlled Thermonuclear Reactions.* New York: Van Nostrand, 1960, chap. 2.

7. R. Carruthers, D. A. Davenport, and J. T. D. Mitchell, "The Economic Generation of Power from Thermonuclear Fusion," Report CLM-R85, UKAEA., Culham Laboratory, Abingdon, Berkshire, U.K., 1967.

8. M. G. Homeyer, "Thermal and Chemical Aspects of the Thermonuclear Blanket Problem," *Tech Rept. No. 435*, MIT, Res. Lab. Elect., June 29, 1965.

9. J. D. Lee, "Some Neutronic Aspects of a D-T Fusion Reactor," in *Proceedings of the Symposium on Thermonuclear Fusion Reactor Design*, M. Kristiansen and M. O. Hagler, Eds., Plasma Laboratory, Department of Electrical Engineering, Texas Tech University, Lubbock, Texas 79409, June 2–5, 1970.

10. D. Steiner, private communication.

11. P. J. Persiani, M. C. Lipinski, and A. J. Hatch, "Survey of Thermonuclear Reactor Parameters," *Proc. Texas Symposium on the Technology of Controlled Thermonuclear Fusion Experiments and the Engineering Aspects of Fusion Reactors*, Univ. Texas at Austin, Nov. 20–22, 1972.

12. Moshe J. Lubin and Arthus P. Fraas, "Fusion by Laser," *Sci. Am.,* 224, 21 (1971).

13. James Benford and Gregory Benford, "Intense Electron Beams—A Fusion Match?" *New Scientist,* 55, 514 (1972).

14. Edward Teller, "A Future ICE (Thermonuclear, That Is!)," *IEEE Spectrum,* 10, 60 (1973).

15. F. Winterberg, "Initiation of Thermonuclear Reactions by High-Current Electron Beams," *Nuclear Fusion,* 12, 353 (1972).

16. G. Yonas, "Engineering Concepts in Fusion," *Proc. 16th Annual ASME Symp. on Energy Alternatives*, Albuquerque, New Mexico, February 1976.

17. William D. Metz, "Fusion Research (I): What Is the Program Buying the Country?" *Science* 192, 1320 (1976).

18. William D. Metz, "Fusion Research (II): Detailed Reactor Studies Identify More Problems," *Science* 193, 38 (1976).

19. William D. Metz, "Fusion Research (III): New Interest in Fusion-Assisted Breeders," *Science* 193, 307 (1976).

20. S. O. Dean, et al., "Status and Objectives of Tokamak Systems for Fusion Research," *USERDA Report WASH-1295, 1974* (available from the Superintendent of Documents, U.S. Government Printing Office).

21. H. P. Furth, "Tokamak Research," *Nuclear Fusion* 15, 487 (1975).

22. George H. Miley, *Fusion Energy Conversion* (American Nuclear Society, Hinsdale, Illinois, 1976), chap. 3.

23. Francis F. Chen, *Introduction to Plasma Physics* (Plenum Press, New York, 1974), pp. 140–143.

24. William M. Tang, "Stability Theory in Tokamaks," in *Lectures From the Argonne Faculty Institute on Fusion Plasmas*, George H. Miley, Ed., Center for Educational Affairs, Argonne National Laboratory, Argonne, Illinois 60439 (1975).

25. E. H. Holt and R. E. Haskell, *Foundations of Plasma Dynamics*, (Macmillan, New York, 1965) p. 268.

26. H. P. Furth, "High-Beta Plasma Confinement in the Tokamak Configuration," *Comments on Plasma Physics and Controlled Fusion* 2, 119 (1976).

27. R. F. Post, "Controlled Fusion Research and High Temperature Plasmas," in *Annual Review of Nuclear Science,* Vol. 20, Emilio Segre, Ed. (Annual Reviews, Inc., Palo Alto, California, 1970) pp. 509–588.

28. R. C. Davidson and J. P. Freidberg, "Review of Toroidal Theta-Pinch Theory," *Proc. 3rd Topical Conference on Pulsed High Beta Plasmas,* Culham Laboratory, Abingdon, Oxfordshire, U.K., September 9–12, 1975.

29. Keith A. Brueckner and Siebe Jorna, "Laser-Driven Fusion," *Rev. Mod. Phys.* 46, 325 (1974).

30. J. Reece Roth, "Alternative Approaches to Fusion," *NASA Technical Memorandum NASA TMX-73429*, Lewis Research Center, Cleveland, Ohio, May 1976.

31. J. Reece Roth, "Preliminary Scaling Laws for Plasma Current, Ion Kinetic Temperature, and Plasma Number Density in the NASA Lewis Bumpy Torus Plasma," *NASA Technical Memorandum NASA TMX-73434*, Lewis Research Center, Cleveland, Ohio, May 1976.

32. H. H. Fleischmann and T. Kammash, "System Analysis of the Ion-Ring-Compressor Approach to Fusion," *Nuclear Fusion* 15, 1143 (1975).

33. R. J. Briggs, G. D. Porter, B. W. Stallard, J. Taskar, and P. B. Weiss, "Efficient Trapping of High-Level E-Layers in a Strong Toroidal Field," *Phys. Fluids* 16, 1934 (1973).

34. H. A. Davis, R. A. Meyer, and H. H. Fleischmann, "Generation of Strong Relativistic Electron Rings with Millisecond Lifetimes," *Phys. Rev. Letters* 37, 542 (1976).

35. D. C. Robinson, "High-β Diffuse Pinch Configurations," *Plasma Physics* 13, 439 (1970).

36. Glasstone and Lovberg, *Controlled Thermonuclear Reactions*, chap. 8.

37. H. P. Furth in *Controlled Thermonuclear Research, Hearings before the Subcommittee on Research, Development, and Radiation of the Joint Committee on Atomic Energy*, Congress of the United States, Ninety-Second Congress, First Session on The Current Status of the Thermonuclear Research Program in the United States, November 10 and 11, 1971 (available from the U.S. Government Printing Office).

38. G. Gourdon, D. Marty, E. K. Maschke, and J. Touche, "The Torsatron without Toroidal Field Coils as a Solution of the Divertor Problem," *Nuclear Fusion* 11, 161 (1971).

39. A. A. Boozer and M. A. Levine, "Particle Trapping in Magnetic Line Cusps," *Phys. Rev. Letters* 39, 1287 (1973).

40. J. Marshall and F. L. Ribe, "Interim Draft of Alternate Concept Evaluation," Los Alamos Scientific Laboratory, July 1976 (unpublished).

41. K. H. Brown, H. R. P. Ferguson, J. H. Gardner, L. V. Knight, and H. M. Nelson, "The Topolotron: A High Beta Device with Topological Stability," *Proc. Second IEEE Conf. on Plasma Science*, Ann Arbor, Mich., May 14–16, 1975.

42. A. Y. Wong, Y. Nakamura, B. H. Quon, and J. M. Dawson, "Surface Magnetic Confinement," *Report PPG-215*, Plasma Physics Group, University of California at Los Angeles, March 1975.

43. "The Application of High Current Relativistic Electron Beams in Controlled Thermonuclear Research," *Report WASH-1286*, Division of Controlled Thermonuclear Research, Research Program, U.S. Atomic Energy Commission, 1973.

44. "Electron Beam Fusion Program," Sandia Laboratories, Albuquerque, New Mexico, March 11, 1976.

45. G. Decker and R. Wienecke, "Plasma Focus Devices," *Proc. 12th International Conference on Phenomena in Ionized Gases*, Technische Hogeschool te Eindhoven, Eindhoven, Netherlands, August 18–22, 1975.

46. V. S. Imshennik, N. V. Fillipov, T. I. Fillipova, "Similarity Theory and Increased Neutron Yield in a Plasma Focus," *Nuclear Fusion 13*, 929 (1973).

47. H. F. Vogel, M. Brennan, W. G. Dase, K. M. Tolk, and W. F. Weldon, "Energy Storage and Transfer with Homopolar Machine for a Linear Theta-Pinch Hybrid Reactor," Report LA-6174, Los Alamos Scientific Laboratory, April 1976.

48. Michael J. Schaffer, "Self-Driving Liner Compression Systems," *Proc. IEEE International Pulsed Power Conference*, Texas Tech University, Lubbock, Texas, November 9–11, 1976.

49. Loren C. Steinhauer, "A Feasibility Study of a Linear Laser Heated Solenoid Fusion Reactor," *Report EPRI ER–171*, Electric Power Research Institute, Palo Alto, California, February 1976.

50. M. Kristiansen and M. O. Hagler, "Laser Heating of Magnetized Plasmas," *Nuclear Fusion 16*, 999 (1976).

Annotated Bibliography

Since the United States program in controlled fusion was declassified in the 1950s, a number of useful books about controlled fusion and plasma physics have become available. We list here only a few of these which we have found to be interesting and useful. The level of presentation in most of these books is more advanced than that we have used.

1. Amasa S. Bishop, *Project Sherwood* (Addison-Wesley, Reading, Massachusetts, 1958).

This book outlines in a fascinating way the early work in controlled fusion. Watch for the genesis of the concept of minimum-B and the Surmac. Don't miss the unusual approach to fusion described in Appendix II!

2. Lev A. Arzimovich, *Elementary Plasma Physics* (Blaisdell, New York, 1965).

This book represents a reasonably successful attempt to liberate the basic concepts of elementary plasma physics from the mathematics which besets them. Unfortunately, much plasma physics has developed after this book was written and is not included. Even so, it can be very useful to the novice in controlled fusion.

3. Samuel Glasstone and Ralph H. Lovberg, *Controlled Thermonuclear Reactions* (Van Nostrand, New York, 1960).

This aging volume seems irreplaceable as a repository of simple physical treatments of the conditions for thermonuclear reactions, certain approaches to fusion and plasma diagnostics, and energy losses and scaling laws. It has therefore remained useful despite the fact that such longtime favorites as Tokamaks and toroidal theta pinches, not to mention laser fusion, are not even mentioned. (Notice paragraphs 8.87 and 11.8, however.) The biggest problem for controlled fusion tyros is in sorting out the outdated material.

4. Terry Kammash, *Fusion Reactor Physics* (Ann Arbor Science, Ann Arbor, Michigan, 1975).

This book offers a more up-to-date and detailed treatment of fusion reactors than Glasstone and Lovberg.

5. George H. Miley, *Fusion Energy Conversion* (American Nuclear Society, Hinsdale, Illinois, 1976).

The title fails to suggest the wealth of material discussed. Browse through this one.

6. Francis F. Chen, *Introduction to Plasma Physics* (Plenum, New York, 1974).

This book presents simple, clear, discussions of an impressive variety of topics in plasma physics and controlled fusion at an elementary level.

The beginner in fusion can hardly do better for an introductory plasma text.

7. Nicholas A. Krall and Alvin W. Trivelpiece, *Principles of Plasma Physics* (McGraw-Hill Book Co., New York, 1973).

This text comes closer than any to bridging the gap between the plasma physics of most textbooks and the plasma physics of the literature.

Index

Index

adiabatic invariant, 99
alpha particles, 3, 33, 48, 49, 118, 145, 176
ambipolar diffusion, 100
ambipolar potential, 100, 123

Astron, 123, 151–153; e-layer, 151–153, 175; field reversal, 152; p-layer, 152–153
banana orbits, 76, 80, 81, 84, 100
baseball coil, 111
beams: electron, 35, 43, 44, 151–153; ion, 165, 166; neutral, 34, 47, 90, 94, 95, 114, 115
beta, 46, 57, 63–67, 69, 87, 89, 90, 98, 114, 122, 134, 135, 149, 151, 153, 156, 158, 159, 171, 175; maximum for tokamaks, 66, 90, 98; poloidal, 65, 67, 86, 87, 89, 101
Bohm diffusion. *See* diffusion, anomalous
bootstrap current, 97
breakeven, 49, 50, 53, 54, 118, 134, 137; temperature, 21, 27
bremsstrahlung. *See* radiation
bremsstrahlung, inverse 170, 175, 176
bumpy torus, 121, 149–151, 161
burn-up fraction, 138–140, 143, 145, 146

circulating power, 45, 49, 94, 117, 118, 120, 122, 137
collisions, 7, 13, 34, 69, 76, 79–85, 97, 100, 114, 148, 176; frequencies, 80–82, 87, 97–99, 112, 114, 115
compression, 35, 43, 44, 47, 134, 139, 140, 152, 153, 164–174; adiabatic, 95, 125, 174; pellet, 139–141, 163

confinement time, 28–30, 52, 85, 86, 122, 143, 155, 166, 171, 174; energy, 101, 122; particle, 101, 114, 122, 149
Coulomb scattering, 7, 80
curvature, magnetic field, 23, 24, 65, 89, 90, 111, 134, 149, 156, 158, 159
cusps, 158, 159; stuffed, 156, 158, 175
cyclotron frequency, 71, 81

diffusion, 97, 121, 153, 156, 167; ambipolar, 100; anomalous (Bohm, enhanced), 15, 16, 83, 85, 125, 149, 155; classical, 80, 149; coefficients, 85–87, 100; dissipative trapped ion mode, 85; neoclassical, 81; random walk model, 79
direct energy conversion, 47, 48, 115–120, 127, 130–132
divertor, 35, 48, 92, 93, 120, 175
drifts, particle, 9, 69–71, 149, 158; $\mathbf{E} \times \mathbf{B}$, 19, 149, 158; grad B, 19
drift waves, 100, 132, 134, 155
dynamic stabilization, 15, 46

electron beam pellet fusion, 35, 43, 162, 163, 166
electron beams. *See* beams, electron
end loss, 7–10, 59, 114, 115, 120, 121, 127, 134, 171, 174
energy storage: capacitive, 129, 130; homopolar generator, 175; inductive, 129, 130, 132, 168
equilibrium: plasma and magnetic field, 20, 63, 97, 149, 153

feedback stabilization, 15, 16, 46, 127

185

About the Authors

Marion Hagler is professor of electrical engineering at Texas Tech University. He received the Ph. D. degree in electrical engineering from the University of Texas at Austin in 1967. He began to participate in fusion research in 1963. Dr. Hagler's present research interests include radio frequency heating of plasmas, laser heating of plasmas, plasma-solid interaction, and coherent optical systems. He has served as coorganizer of several conferences related to controlled fusion, among them the Symposium on Fusion Reactor Design in 1970, and the First and Second Topical Conferences on RF Plasma Heating in 1972 and 1974, respectively.

Magne Kristiansen was awarded the Ph.D. in electrical engineering by the University of Texas at Austin in 1967. He then joined the faculty of electrical engineering at Texas Tech University, where he is now a professor. Dr. Kristiansen is especially interested in using lasers and radio frequency sources to heat plasmas, in producing radiation by plasma-solid interaction, and in developing high-voltage high-current pulsed power techniques. Since he entered the fusion field in 1961, Dr. Kristiansen has organized numerous conferences related to fusion, has served on fusion-related advisory panels for federal agencies, and has worked at Los Alamos Scientific Laboratory as a visiting staff member on several occasions. In 1975, he was awarded a NATO Senior Fellowship for fusion research at the Max-Planck-Institut für Plasmaphysik, Garching bei München, FRG.